# PROFESSIONAL SHEET METAL FABRICATION

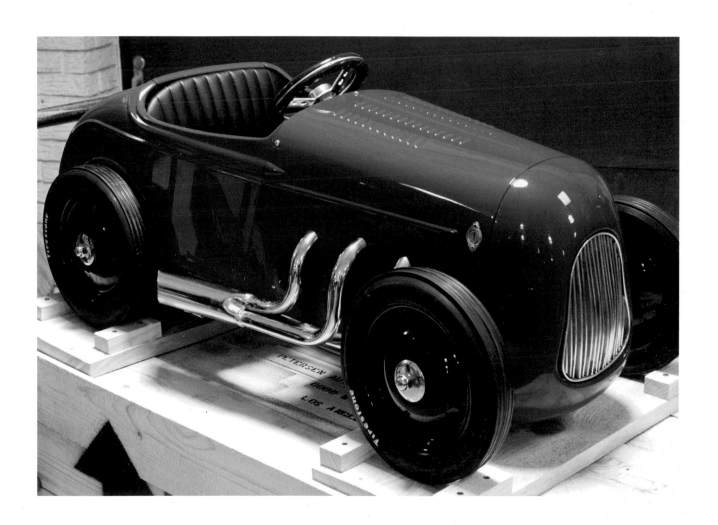

By Ed Barr

motorbooks

First published in 2013 by Motorbooks, an imprint of
Quarto Publishing Group USA Inc., 400 First Avenue
North, Suite 400, Minneapolis, MN 55401 USA

Motorbooks titles are also available at discounts in bulk
quantity for industrial or sales-promotional use. For details
write to Special Sales Manager at Quarto Publishing Group
USA Inc., 400 First Avenue North, Suite 400, Minneapolis,
MN 55401 USA.

To find out more about our books, visit us online at
www.motorbooks.com.

Library of Congress Cataloging-in-Publication Data

Barr, Ed.
   Professional sheet metal fabrication / by Ed Barr.
      pages cm
   Summary: "Ed Barr's Professional Sheet Metal Fabrication
is a comprehensive guide to workshop shaping and
manipulation of sheet metal for both automotive projects
and other metalworking endeavors. Barr is a certified
professional who teaches metalwork at McPherson College,
the only licensed sheet metal fabrication program in the
United States"-- Provided by publisher.
   ISBN 978-0-7603-4492-7 (pbk.)
   1.  Automobiles--Bodies. 2.  Sheet-metal work.  I. Title.
   TL255.8.B37 2013
   629.2'60288--dc23
                                        2012044854

Editor: Jordan Wiklund

Design Manager: Brad Springer

Designer: Danielle Smith

Printed in China

10 9 8 7 6 5

# Contents

Introduction . . . . . . . . . . . . . . . . . . . . . . . . . . . . . . . . . . . . . . . . . . . . . . . . . . 4

CHAPTER 1: **Getting Started** . . . . . . . . . . . . . . . . . . . . . . . . . . . . . . . . . . . 5

CHAPTER 2: **Special Techniques for Welding Sheet Metal** . . . . . . . . . . . . . . . 24

CHAPTER 3: **Brazing, Soldering, and Riveting** . . . . . . . . . . . . . . . . . . . . . . 62

CHAPTER 4: **Cutting Sheet Metal with the Oxyacetylene Torch
and Plasma Cutter** . . . . . . . . . . . . . . . . . . . . . . . . . . . . . . . . . 79

CHAPTER 5: **Beginning Sheet Metal Shaping** . . . . . . . . . . . . . . . . . . . . . . . . 94

CHAPTER 6: **The Small Gas Tank Project** . . . . . . . . . . . . . . . . . . . . . . . . . . . 150

CHAPTER 7: **Advanced Sheet Metal Shaping** . . . . . . . . . . . . . . . . . . . . . . . . 160

CHAPTER 8: **Building a Fender from Concept to Completion** . . . . . . . . . . . . 187

CHAPTER 9: **Making It Beautiful: Straightening, Grinding,
and Surface Finishing** . . . . . . . . . . . . . . . . . . . . . . . . . . . . . . . 211

CHAPTER 10: **Building a Custom Pedal Car** . . . . . . . . . . . . . . . . . . . . . . . . . . 244

CHAPTER 11: **Floorpans, Rocker Panels, and Rear Quarter Panels** . . . . . . . . 255

CHAPTER 12: **Repairing Doors** . . . . . . . . . . . . . . . . . . . . . . . . . . . . . . . . . . . . 270

CHAPTER 13: **Repairing Fenders, Hoods, and Trunk Lids** . . . . . . . . . . . . . . . . 280

**Resources & Bibliography** . . . . . . . . . . . . . . . . . . . . . . . . . . . . . . . . . . . . . . . 301

**Index** . . . . . . . . . . . . . . . . . . . . . . . . . . . . . . . . . . . . . . . . . . . . . . . . . . . . . . . . 302

# Introduction

This book is for people who love old cars and who want to make their old car's sheet metal better than new with their two hands. The practice of restoring antique cars is a complex, time-consuming, yet incredibly rewarding experience. It taxes your patience and financial resources, but for some of us it is as essential as oxygen. As with all the different facets of auto restoration—soft trim, paint, mechanical work, and more—sheet metal repair provides the opportunity to transform the quality of something from extremely bad to extremely good or at least in much better condition than it was. Sheet metal fabrication allows you to make something beautiful out of thin air. Whether you are patching a small rust hole or making an entire fender, at the end of the day (or at least at the end of most days), you will feel good about having made tangible, measurable progress.

For people with little experience in shaping metal, the idea of making panels or parts of panels from scratch is more than a little daunting. Shaping sheet metal is intimidating only because the logic behind it isn't immediately apparent. Most car folk understand how a flywheel turns because it is attached to a crankshaft, which turns because pistons push on it in a certain way, and so forth. Sheet metal seems to have a mind of its own, however—you hit it in one spot and it moves as expected, but another part of it may bend or change shape as well. Fortunately, working with sheet metal is neither black magic nor rocket science (though certainly some knowledge of science is important). If you were to ask a group of experienced metal shapers what it takes to succeed in their field, I'm sure it would come down to thoughtful persistence and a knowledge of materials; if at first you don't succeed, take a different approach and try again. Sometimes making something out of sheet metal will take much, much longer than you anticipate, but in my experience effort always pays off in this discipline.

In this book I present a number of techniques for making and repairing automotive sheet metal components. As your skills increase, think of ways to solve progressively difficult problems by combining multiple processes. Many projects seem impossible to carry out unless you break them down into smaller tasks that involve one or more techniques. Furthermore, don't be afraid to make mistakes; making one or more false starts is still progress—you've learned a couple of ways *not* to do something. When talking to students, I call false starts the research and development phase.

The organization of this book resembles the sequence I prefer to follow when introducing new students to sheet metal restoration. I begin with joining and cutting processes because they are absolutely essential to making automotive-related objects out of sheet metal, and acquiring welding skill takes time. Starting with welding means that you'll have more of your life ahead of you to practice. I introduce metal-shaping processes next and conclude with a series of hands-on projects intended to illustrate common automotive sheet metal problems and how to solve them. Now let's get to work.

This Kirkham Motorsports Shelby Cobra replica wears machine-pressed panels that were torch welded together and finished out by hand. The flawless finish speaks to the remarkable skill of the craftsmen who built this car and to the need for people with an appreciation and knowledge of fundamental metal-shaping techniques.

# Chapter 1
# Getting Started

## SAFETY AND FASHION SENSE

Although I have reached a point in life where fashion for the sake of being fashionable is no longer a concern, I still make careful wardrobe choices before I enter the shop. Working with sheet metal can be dangerous, but not if you pay attention to the potential hazards involved before you go to work. Minimize welding and grinding hazards by protecting yourself from head to toe. First, wear safety glasses, either with or without corrective lenses as needed. In total defiance of all conventions of coolness, I wear eyeglasses with the little plastic shields on the sides. These glasses embarrass my children when I wear them in public, but most debris enters your eyes from the sides, so taking precautions is essential. Furthermore, if your incredible good looks make leaving your house a nuisance, these eyeglasses insure that members of the opposite sex will go out of their way to avoid you; you'll have more time for working with metal. If you work in a shop

with other people, some of the greatest hazards will be debris generated by co-workers, especially if they will be grinding. In addition, wear long-sleeve shirts, long pants, and leather shoes or boots to block UV rays generated during welding. Use common sense. Espadrille sandals may have worked for Crockett and Tubbs on *Miami Vice*, but they have no place in an area where you'll be welding and cutting. Spiked heels tend to sever the cord to the TIG welder amperage control, so leave those deep in the closet behind the espadrilles. Every sheet metal edge is a potential cutting edge, so keep sheet metal stored away in a rack when not in use. Wear leather gloves when handling new or freshly cut sheet metal.

When working around old cars, be mindful of the threat they pose as falling objects, as containers for flammable liquids like fuel and brake fluid, and as electrocution hazards. Disconnect the car's battery or remove it altogether. Eliminate

This fender seems complicated, but it is made up of eight individually shaped panels that were less intimidating when conquered one at a time following John Glover's classic video on the subject.

Our 1954 Plymouth Belvedere is the master of disaster. Extensive preparations were necessary in the form of added structural support before work on it could begin because the body was so rusty.

known dangers. Although you have no doubt seen some slack-jawed knucklehead support a car with a jack alone, don't risk it, especially when you consider the proliferation of poor-quality jacks in auto parts stores in recent years.

## MAKE A PLAN

One of my colleagues once said that failing to make a plan was planning to fail. I could not agree more, especially considering the cost in time and materials for making something out of metal. Starting out, you'll probably make some scrap, and then you'll go overboard with planning. Eventually you'll find the right balance between obsessive preparation and action. If possible, start with small projects to build experience and confidence. When restoring an automobile, choose your projects in a sequence that suits your skill level. Think several steps ahead of where you are, and keep notes of the best possible sequence of events for your project. All professionals develop schemes for planning and organization, or they do not stay in business very long. Planning your projects will minimize time and materials wasted as well as cutting down on the frustration that accompanies hasty, foolish mistakes. Fortunately, the mental clarity that follows expensive, boneheaded mistakes is particularly acute and goes a long way toward keeping some of us from repeating our errors—at least some of the time.

With projects of any size, consider keeping a clipboard and legal pad handy for writing down notes and making sketches. If you are a nonlinear thinker, make lists with bullet points and then go back and number the items in the proper sequence so that you will be able to work efficiently. Making things out of metal is difficult enough without the added aggravation of having to retrace your steps every time you get interrupted or, worse, having to redo work that was done in the wrong sequence. If you find yourself juggling several projects simultaneously, you will notice an increase in your productivity and work satisfaction resulting from simple task lists that give you direction and focus. If the project at hand can be done in an afternoon, perhaps elaborate preparations will not be needed. But if there is any danger of interruption, at least scribble a note on a piece of masking tape regarding where you were and what you were going to do next. This practice will give you a head start when you return to the project.

Whether you are striving to increase the quality of your restoration or simply want to avoid future aggravation, cultivate a forward-looking approach to your work. When faced with a rusty panel that needs repair or replacement, for instance, your first instinct may be to cut out the damaged part. Sometimes removing the old metal is the first logical step toward repair, but not always. Instead, consider leaving the deteriorated piece in place as long as possible to allow you to check the fit of your patch or replacement panel.

Before launching into any sheet metal repair, consider the scope of the repair. Sheet metal repairs tend to snowball because more damage is revealed as you peel away each crusty layer of metal. Is your car a driver or a full-blown restoration?

I have had to train myself to think specifically in these terms to prevent every car I work on from becoming a perpetual project. Decide early on what your goal for your car is and stick to it. The longer a vehicle is unusable, the greater its chances of remaining that way. If the project requires extensive metal work and you are committed to seeing it through, I hope the suggestions that follow prove helpful.

For major, potentially structural, repairs, consider the ramifications of each step in the repair process—make a chronological to-do list and keep it with the car. If you plan to use aftermarket repair panels, have them in hand before you begin. You may not be able to assess their quality until you actually try to fit them on the car, but at least you won't cut away a section of metal in anticipation of receiving a part, only to find that it has been discontinued.

Assess how well your doors, hood, fenders, trunk lid, and other panels fit before you change anything about your vehicle. If your car has a frame, are there decayed mounting points, frame pads, or insulating blocks that are negatively influencing panel fitment? Likewise, if the car has unitized construction, is it sagging or damaged in any way? Uneven tire wear or peculiar handling characteristics might be signs of misalignment in the body that you need to correct before cutting and welding. If the rear passenger door has never quite fit right, for example, now is the time to find out why. Perhaps the hinges are bad, but maybe the bottom of the pillar on which the hinges are mounted has rusted away. Once you resolve the major structural and alignment problems,

take whatever measures are necessary to maintain your solid foundation from now on. Typically, those measures will involve welding in structural supports to ensure that nothing moves while you make repairs.

For floorpan, rocker panel, and sill repairs, a quick and easy way to maintain structural stability is to weld one or two pieces of square tubing across the inside of the door opening, depending upon how suspect your rocker and floor are. I have included a photograph of a 1954 Plymouth Belvedere that is as extreme a case of structural decay as a person is likely to encounter. Notice how every structural member of the body is tied to an interlocking, triangulated framework of welded tubing. Hopefully, you won't need support that is so extreme, but the Belvedere should offer ideas of how to fortify your vehicle, depending on what is required. As you add bracing, consider access to carry out the repair and ease of removal once the work is done.

How will you support the car while you work? Rotisseries are nice, but a lot of good restorations have been done without them, so do not feel that they are essential. Depending on the design of the car, rotisseries can be a hassle to attach, and they are undeniably a hassle to store when not in use. Tall, extra-heavy-duty jack stands are nice—quick to set up and take down and easy to make level front to back and side to side. If you make a repair that may influence alignment, such as a leaf spring mount, for example, level your vehicle so you will have an objective reference point to get everything back where it belongs.

Even if your vehicle is supported on a rotisserie, make sure its body is inherently stable enough to keep everything in alignment. We added two additional temporary braces inside this Ford Mustang trunk because the bumper mounts alone did not seem substantial enough to do the job.

## GETTING ORGANIZED

Few experiences are as frustrating as looking forward to spending some quality time with your project car only to end up wasting time looking for tools and parts. As a hobbyist, one's natural tendency is to want to use every available second of free time working, with the result that you leave things in process. Parts and tools tend to land on whatever surface makes itself available at the moment. Meanwhile, work, home repairs, your overgrown lawn, and other pressing concerns intervene. By the time you get back to work, weeks may have elapsed, tools may have been used on other unrelated projects, and parts may have been hastily pushed aside and become mixed together. Such disarray makes completing the project even more difficult and frustrating. Instead, force yourself to be organized in this one area of your life. You will be amazed how much easier it is to make progress when things are always where you expect them to be.

Documentation is a critical part of being organized. Automobiles have literally thousands of fasteners, clips, and sundry parts. Inevitably, you will at least partially disassemble your project to make repairs. Have a system to keep track of everything that comes off in order to get everything back together successfully. Take plenty of pictures and label the pieces as you go, both in the pictures and in the containers in which you store them. I like to write on masking tape and take pictures of parts before I take them off the car so that the label appears in the picture. You can also write on things with chalk, a paint pen, or a permanent marker so that the labels are legible in photographs. Resealable zipper storage bags are handy for storing fasteners, but put a label in the bag as well as labeling the bag itself; oil and grease will remove permanent ink from the outside of bags. Store the bags in cardboard boxes with the parts they accompany, and label the box on two adjacent sides so that it will be easy to spot on a shelf.

If you can force yourself to make a final decision where each tool should go, tool boards are a great way to keep things organized. When everything has a place, it is easy to put things away.

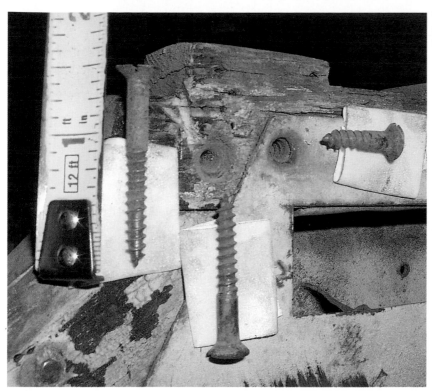

Thorough documentation is essential to putting a restored car back together successfully. Take pictures of how things fit in the process of disassembly.

This is a storage area inside Vintage Restorations in Union Bridge, Maryland, where I was formerly employed. The rigorous organization, cleanliness, and ethical standards practiced there make it a model restoration business. For maximum efficiency, parts were labeled and easy to find.

## PRO SHEET METAL TERMS

**Annealing**—the process of heating metal to a specific point to effect change in the microstructure of the metal, often undertaken to restore ductility to metal that is stiff from work (work-hardened).

**Body hammer**—a hammer specifically designed for automotive body repair. Typically, large versions are "bumping hammers," small versions are "dinging" hammers, and hammers that combine a striking face with a pointed or chisel-face are "combination" hammers.

**Body solder**—solder designed specifically for auto body repair. A common composition is 30% Tin and 70% Lead.

**Buck**—a foam, wood, or wire three-dimensional representation of an anticipated sheet metal project that is used as a guide during shaping.

**Bull's-eye pick**—an archaic hand-held auto body tool that utilizes a long swing arm to remove dents from automotive panels by hitting them from the backside.

**Cleco**—a spring-loaded, temporary rivet. Commonly used in the aircraft industry.

**Cold-shrink**—to shorten and thicken a spot on a sheet metal panel by striking it with a hammer without the use of heat to soften the panel.

**Cold-working**—hammering, bending, or manipulating metal without heat so that the metal is plastically deformed.

**Corking and driving tools**—tools used to shape metal by hitting them with a hammer in the manner of a chisel or punch.

**Dolly block**—a hand-held weighty steel support used to back up or direct hammer blows in auto body sheet metal repair procedures.

**Dzus fastener**—a quarter-turn fastener used to secure sheet metal panels that must be periodically removed.

**Elastic limit**—the point at which sheet metal cannot spring back to its original shape when a force is applied to it. The change in shape past the elastic limit is plastic deformation.

**English Wheel**—a metal shaping tool composed of a large fixed upper steel wheel and a lower, removable anvil wheel. Sheet metal is stretched and therefore shaped by being rolled between the wheels.

**Hammerform**—a sturdy form representing all or part of a finished project. Hammering sheet metal over the form speeds the shaping process.

**Heat-shrink**—to thicken and shorten an area on a sheet metal panel by striking it while it is hot.

**Linear stretching die or hammer**—a striking tool with an elongated, rather than round, face. The elongated contact surface moves more metal adjacent to the long sides of the tool than at the ends.

**On and off dolly**—refers to whether hammer blows are directed against or next to a dolly block during auto body sheet metal repair operations.

**Planish/planishing hammer**—To planish is to smooth by hammering. A planishing hammer is a pneumatic tool used to smooth sheet metal rapidly by repeatedly hammering it between removable dies.

**Shrink**—to compress or shorten sheet metal, typically making it thicker in the process.

**Spoons and slappers**—large, flat striking tools used to smooth sheet metal.

**Stake**—a heavy object against which sheet metal is shaped by hammering, usually mounted on a post and secured in a vise.

**Stretch**—to lengthen sheet metal, typically making it thinner in the process.

**Thumbnail dies**—dies used in various machines to shrink sheet metal by raising up a wrinkle in the metal and subsequently compressing the metal back into itself through hammering.

**T-stake**—a heavy post-mounted object in the shape of a large "T" against which sheet metal is hammered and shaped.

**Tucking tool**—a tool used to create wrinkles or tucks along the edge of a sheet metal panel through twisting. The wrinkles are shrunk by hammering.

**Work-hardening**—the strengthening of metal through cold working.

## TOOLS

Once you have reinvented the new, more-organized you, what tools will you need to realize your sheet metal dreams? First, you want a small assortment of automotive body tools. They are highly developed for correcting damage, and they are great for shaping and smoothing pieces of sheet metal.

Basic automotive hand tools have changed little in design in the last 80 or so years. Quality has changed, however. If ever you get your hands on some Proto or Martin wood-handled body tools, or vintage Fairmount or Plumb tools, you'll notice the difference each time you must handle a sad imitator cast from discarded tin cans and bicycle frames in a land far away.

### Hammers and Dollies

Old body hammers feel good in your hand the moment you pick them up, and old dollies are weighty and dense. Working with nice hand tools is one of the great tactile pleasures of metal working that should not be missed if you can help it, but nice can be pricey if you want nice and new. You may be able to find some usable old body tools on the Internet, at a swap meet, or at an estate auction in the midst of a lot of random tools, however. Best of all, a thin disguise of grime, surface rust, and perhaps a cracked handle may conceal many superb old body tools from the indiscriminating hordes jostling about at your local estate auction or garage sale. If you are reluctant to buy new Martin body tools and are unable to find acceptable old tools, you can make do with whatever you can find. The metal will not care if you hit it with a nice body hammer or your forehead. Eventually, you may find that some of your favorite hammers are ones you've made by regrinding the heads of old ball-peen hammers, though their short reach can be a limiting factor if you need to hammer against the back side of a deeply curved piece of metal.

The type and number of tools you need will depend on the kind of work you intend to do. Fortunately, you can accomplish a lot with very few body tools. A body hammer, a dolly, a block of wood, and a spoon will take you far, and you can add pieces as the need arises. Although the following advice may seem a little odd, I encourage you to see *all* hammers, and many other objects as well, as *potential* body hammers. Short of personal injury and animal cruelty, I encourage creative free thought regarding selecting a tool for shaping metal. In this form of personal expression, the artist cannot be constrained by arbitrary labels like *body hammers*, *ball-peen hammers*, and *bowling balls*. Your tools are here to serve you, the taskmaster, so use them as you wish to achieve the results you want.

Automotive body hammers come in three basic types: bumping hammers, dinging hammers, and combination hammers. A bumping hammer typically has a larger head than does a dinging hammer because its purpose is to bump large dents back into shape. The dinging hammer, therefore, picks up where the bumping hammer leaves off; its job is to remove small dents. The combination hammer usually has a small dinging face opposite a pick. The pick is used for raising small, very specific areas or for covering the back side of your work piece with numerous tiny unsightly and difficult-to-remove dents. I will discuss the pick in a later chapter.

The faces of body hammers are either flat, or almost flat, for dinging on the outside of a crowned panel, or they are curved, for working behind a crowned panel. In the most perfect of worlds, the faces of your hammers should be kept smooth and polished because any imperfection on a hammer or dolly face will be transferred upon impact to the work surface. In the real world, however, this level of quality is only an issue when you are hammering metal directly on a dolly. Again, the kind of work you do will determine whether that degree of attention to detail is necessary. One other hammer face you might see is the so-called shrinking hammer, which has a grid of small squares embossed into the hammer's face. This hammer is an abomination, in my opinion, and should be avoided, though you may be able to buy a used one for cheap and then grind the face smooth. In my experience, this hammer leaves a grisly trail of stretched metal wherever it has

been used. I suppose the original theory behind its design—assuming that there was a theory—was that the squares would upset the metal into the interstices separating the squares each time the hammer struck metal against a dolly. It seems to me, paradoxically, that this upsetting can only take place if the small squares displace, and therefore *stretch*, the metal upon impact. In practice, the net gain of the shrinking hammer is decidedly toward the stretching side of the equation, so beware. Likewise, I would expect the shrinking dolly blocks that are similarly embossed with squares to stretch metal rather than shrink it, but I have not used one. The shrinking hammer I have just described should not be confused with another shrinking hammer that has a round spring-loaded head with grooves radiating out from the center of the face in a spiral. The latter applies a twisting motion upon impact. I have never used one of these hammers and cannot attest to its effectiveness. I will discuss other shrinking methods later in this book in connection with dent repair.

The cruelly misnamed shrinking hammer is as unholy an instrument as has ever cursed the earth. If you can pick one up for cheap, regrind the face and make a useful tool out of it.

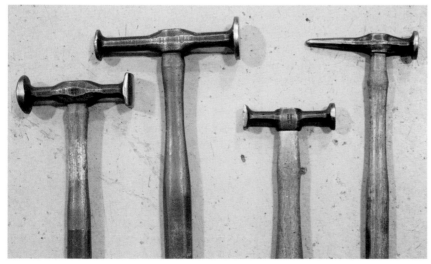

From left: two bumping hammers, a dinging hammer, and a hammer that combines a pick and a dinging head—a combination hammer. The pick is almost never used.

Anything can be a dolly block, and you can never have too many. The standard shapes are the toe and heel, the general purpose, the wedge/comma, and the loaf dolly. If you're extra possessive of your tools, feel free to think of the comma dolly as an apostrophe.

Big spoons are usually bumping spoons. Small spoons are dinging spoons. Odd-shaped spoons can be used as prying tools, as driving tools, or as dollies. The two slender spoons on the right are slapping spoons. They are good for hitting people who use shrinking hammers, among other things.

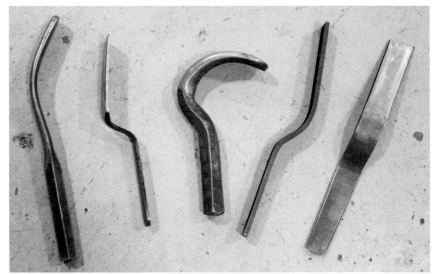

The hammer's counterpart, the dolly block, is placed behind the panel being struck to direct and control the force of the hammer blow. Although you might assume that you simply hold the block behind a dent and commence hammering, this method is not the best way to straighten metal. Although some straightening can be accomplished this way, hammering on dolly, as the technique is called, can also very easily stretch the metal. Instead, more often you will find yourself hammering next to the dolly, or off dolly. To get the most from your dolly or dollies, train yourself early to see every surface of each dolly block as a potential hammering platform. A handful of common dolly shapes have evolved to suit just about every repair situation. Nevertheless, you will feel the need from time to time to call into service and reshape as dollies other random scraps of metal to suit a particular project when one of the standard dolly shapes does not work.

## Spoons and Slappers

Spoons and slappers are the hybrid cousins of hammers and dollies. They can act like prying tools, dollies, or hammers if needed. The family of spoons encompasses a number of broad, flat spatula-shaped hand tools of varying sizes and weights. As prying tools, spoons can reach into crevices to apply pressure to the back sides of dents that cannot be reached by other means. A spoon functions as a dolly when it is slid behind a panel to offer resistance to a hammer blow, either directed next to the spoon or on top of it. By placing a spoon face-down on a piece of metal and hammering on the back side of the spoon, you can spread the force of a hammer blow over a large area without leaving hammer marks. You can also hit metal directly with a spoon as you would with a hammer. The benefit is that you will leave very little evidence of your presence on the work because the hitting surface is so large. Spoons used in this manner may be used to smooth metal to an astonishing degree. As a finishing tool, the "slapping spoon," or simply "slapper," is the refined aristocrat of the spoon family. Because of its large flat contact area and balance, the slapper is often the tool of choice for many professionals for the final phase of straightening. When struck on dolly, the slapper will level groups of subtle dents that would be tedious to resolve one at a time. Some slapping spoons you may come across are endowed with rows of teeth, like a file. I suspect these are another version of the waffle-faced shrinking hammer and shrinking dolly. I do not know if these teeth are intended to shrink the metal in the manner of the so-called shrinking hammer or whether the teeth are intended simply to score the metal to identify where the slapper has struck, thereby allowing you to gauge your progress. I can attest to the ability of these teeth to scar the metal, however. Should you find one of these slappers, I would recommend grinding the teeth smooth on the hitting face and possibly the handle, too, if you find them to be uncomfortable in your hand. I have experimented with both the serrated slapper and the smooth slapper, and I cannot discern any enhanced performance from the serrations, only marks.

## Corking and Driving Tools

Another group of automotive body tools that may be helpful in general sheet metal straightening and shaping are the corking and driving tools. All of these tools are meant to be hit with a hammer to apply force to a panel. A corking tool is used to define a fold, bend a flange, or move metal in a clearly defined area that is either too confined or too risky to hit with a hammer, or the user wants a tool that is softer than the metal so as not to leave hammer marks and facilitate cold shrinking. Driving tools are usually designed to remove dents in areas that cannot be reached with a hammer. Corking and driving tools can be something as simple as a large chisel with a slightly modified point, a piece of hardwood, or a specially forged and curved implement for straightening automotive fender lips, for example. There is probably no need to start amassing a large number of these tools because they can be somewhat specialized, but you should keep their existence in the back of your mind to inspire you to build the tool when you need it.

## Prying Tools

Prying tools facilitate access to tight spaces, such as between the exterior panels of an automobile and their support structure. Prying tools come in handy, too, if ever you need to remove dents from the inside of a container, such as a motorcycle gas tank. In addition, prying tools work well for retrieving fallen sockets that have rolled into nearly inaccessible spaces. As with the corking and driving tools, commit the photograph of them to memory for future use in case you need to make a facsimile someday.

The so-called bull's-eye pick is another specialized automotive body tool worth investigating. The *bull's-eye* part in its name is certainly an exaggeration created years ago by an optimistic advertiser. Nevertheless, this oft-maligned instrument can work well for occasional use, but several factors have to be right. First, the joint in the middle of the tool cannot have any excess play side-to-side. Second, use common sense when selecting the best removable tip for the hammering end. A blunt tip is almost always better than a sharp one, and a hard steel tip is going to be unforgiving on

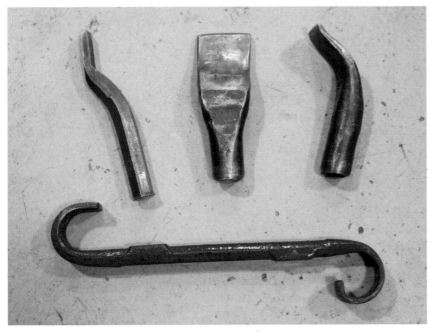

Corking and driving tools are meant to be hammered on their ends like a chisel. The elongated tool at bottom is for straightening fender lips. The small flat spots on the shaft are sites for hammering.

Prying tools are for getting into virtually inaccessible places and for scratching hard-to-reach itches.

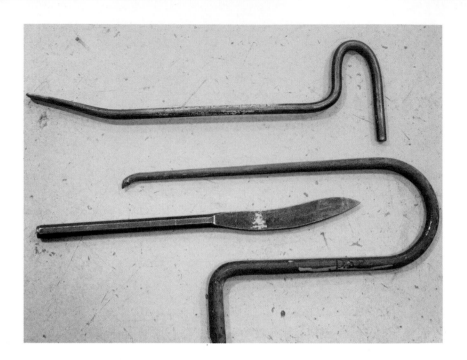

softer metals. You can make tips by grinding the head of a bolt and screwing the tool in place of the normal tip. Third, do not expect to *see* the metal move as you operate the tool. If you can see the metal moving, you are probably hitting the metal too hard. Fourth, something about the elliptical nature of the swinging arm tends, in my opinion, to accelerate the tip toward the work, so beware. Fifth, try mashing the target end of the tool against the work piece with your thumb so that your digit is resting on the tool and the work at all times. If not using your thumb, at least hold the target end of the bull's-eye pick tightly against the work. Pressing the tool to the work while you operate it keeps your blows on target and gives you direct feedback regarding the force of your swings. Sixth, pray that the stars are in their proper alignment.

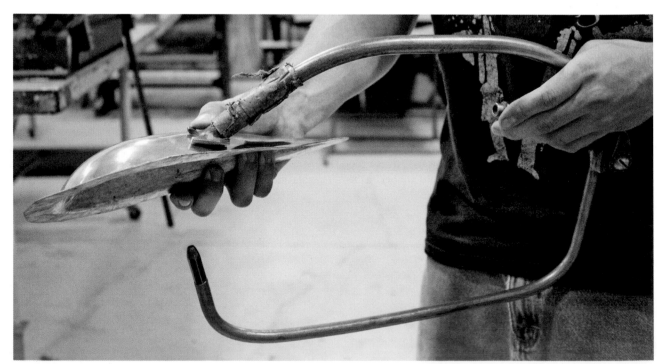

The bull's-eye pick works acceptably if the tool is in good shape, the tip isn't too sharp, and the work isn't hit too hard.

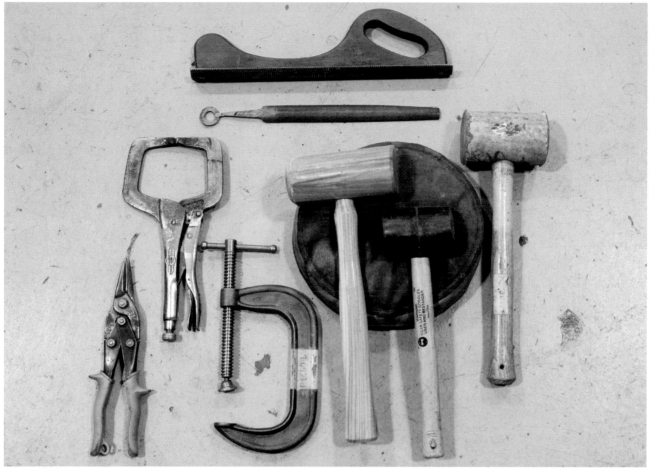

A few essentials include a shot bag, aviation snips, a wood mallet, a rubber mallet, a rawhide mallet, a bastard file, a vixen file (very coarse), and as many Vise-Grips and C-clamps as you can lay your hands on.

I disliked the bull's-eye pick when I first used it, but have grown fond of it over time for use on small dents. This tool seems to work acceptably for dents that are hard to pinpoint from the back side of a panel and yet are a little small for hitting with a hammer or slapper on a dolly.

## Miscellaneous Tools

You will want some other hand tools right away, such as a good pair of aviation snips, a wood mallet, a hard rubber mallet, some C-clamps, a sharp bastard file with flat and half-round sides, and as many authentic Vise-Grips as you can afford. Plastic mallets don't have much mass, so they are not very efficient tools compared with steel hammers or a wood mallet. A dense rubber mallet, on the other hand, works surprisingly well for bumping out large dents. Best of all, rubber mallets are some of the least expensive tools you are likely to come across. Rawhide mallets do not have much mass, but they can be used to cold shrink metal over a hard surface, a technique I will cover in another chapter. If you get only one semisoft hammer, the wood mallet is probably the most useful because the face can be easily shaped to hit the back side of curved panels. A leather bag filled with lead shot or sand is extremely useful for making the most out of your hammers, both for shaping and for straightening bent metal. These bags can be purchased for a reasonable price, but I'll show you how to make one at the end of this chapter if you are so inclined.

The aforementioned tools are the essentials. As funds allow, consider adding a sheet metal brake for bending and some kind of shear. Sheet metal can be clamped against a table and hammered over at an angle, of course, but you will get faster and cleaner results in a brake. A finger or leaf brake is the most versatile because it has removable fingers, or leaves, that allow you to make multiple bends without interference. For cutting metal, you can get by with a jigsaw, band saw, nibbler, or pneumatic cut-off wheel at least for a while, but eventually you will become aggravated by the inability to get the exact clean cut you want in one step every time. If you can afford only one type of shear, keep in mind that a throatless Beverly shear will cut along a straight or curving line, whereas a stomp shear cuts only straight. The Beverly shear is somewhat limited by the size of panel you can guide through it with one hand, however.

A throatless shear, such as the legendary Beverly shear, can cut sheet metal along sharp curves or straight lines.

A sheet metal brake is not essential, but even a small, relatively inexpensive one will be used daily. A finger or leaf brake is the most flexible because the removable sections allow you to make multiple bends, as here.

For cleaning up the edges of sheet metal and for finishing welds, a good file is hard to beat. Bastard files are good for general use, and coarse vixen files are good for checking panels for flatness, filing lead body solder, and removing tall weld beads, especially on aluminum. Some kind of hand-held grinder with about a 5-inch abrasive disc is handy for shaving down the edges of panels and removing extra metal from MIG weld beads. Select a tool with interchangeable backing discs so that you can use a firm back for grinding and a soft pad for finish sanding. For finishing metal to a high degree, a sander with random orbit capability allows you to smooth a surface without leaving any grinder-induced flat spots. A small-angle grinder with a 3-inch Roloc abrasive disc is useful when extra control is needed, but it tends to be overused, and the small abrasive surface leaves a ground surface more uneven than

does a larger-diameter disc. Cut-off tools are incredibly useful for their low cost, and you can call them into service for other grinding duties if you install a thicker grinding wheel in place of the cut-off disc. Electric tools tend to be quieter, more powerful, and more efficient than pneumatic ones, and they don't require a massive air compressor, but the convenience of quick-change air tools makes them a popular choice.

Car folk are well acquainted with the Eastwood Company and its vast assortment of products for the restoration industry. Two Eastwood metal working tools that are well worth purchasing are the pneumatic combination hole punch and flanging tool and the hand-operated shrinker/stretcher. I will discuss the use of these tools later in the book, but I want to mention them now so that you can put them into your budget. If you anticipate the need to simulate factory

If you decide to use air tools, some great ones to have are the following, from left: a pneumatic cut-off tool, the same tool with a weld-grinding wheel installed, an angle grinder with a 3-inch 3M Roloc grinding disc, a 5-inch disc grinder, a 6-inch soft pad sander, and a dual-action palm sander with a soft 6-inch disc. Buy cheap air tools. They are a far better value than the expensive ones—honest.

The Eastwood shrinker/stretcher and the combination hole punch/flanger are essential, or at least similar tools by a different maker are. If you spend much time in the shop, you will use these, especially the shrinker, daily.

resistance welds with a MIG welder, the hole punch will pay for itself the first time you use it. The hole punch allows you to place perfect holes around the edges of a patch panel in a tiny fraction of the time it would take to drill them. The flanger puts a small offset along the edge of the panel similar to that created by a bead roller. The shrinker/stretcher machine allows you to influence the shape of a panel dramatically by changing the thickness and length of the metal along the panel's edge. You will see the shrinker/stretcher used countless times on the projects in this book.

## Bead Roller

Metal shapers have exploited the tendency of metal to work-harden for centuries. Work-hardening—the stiffening of metal through cold working—strengthens metal without adding weight or changing its composition. In everything from armor to Zeppelins, work-hardened metal has been utilized for structural and decorative purposes. In the automobile, for example, pressed beads make panels more resistant to impact damage and vibration, which in turn reduces noise and enhances riding comfort. The stiffness of the automobile body improves handling and safety and increases the longevity of the car. Furthermore, beads just look neat. See, for example, the Chrysler Crossfire and Ford Flex, two modern automobiles with extensive beading over their bodies. Anyone dabbling with old cars inevitably will need to know how to make and repair beaded panels. Fortunately, a basic bead rolling machine is neither complex nor expensive, and it is easy to use. Make sure your roller dies are the correct shape for your application, align them properly, adjust the squeeze the dies exert on your metal, and you are ready to roll a bead. For some dies, the spacing between them is variable and changes the appearance of the final bead, as does the pressure, but that is about as complex as it gets.

A decent bead roller can be had for under $200. It is useful for running beads in panels and floor pans and is handy for putting flanges on the edges of panels.

Because the panel is only reading the contact patch—it has no idea what the rest of the anvil wheel looks like—installing the highest crowned anvil wheel *will not* directly deliver a panel with a shape corresponding to that anvil. The shortest route to frustration with the English wheel is to install the most severely crowned anvil wheel, crank up the pressure, and roll out a panel resembling a long fluted banana. When starting out, stick with the flatter wheels. Use the more rounded wheels when the work piece you have has so much shape that no other wheel will fit or when you want to stretch a very specific area. For more instruction on the wheel than can be included in a single book such as this, seek out a video by John Glover. I suspect that a large portion of the correct information in the world about the English wheel emanated from him originally.

## Slip Roll

The slip roll is not a machine most car enthusiasts need to restore their projects, but I want to demonstrate how it works in the event some readers have a use for it. Slip rolls have two parallel rollers that squeeze the sheet metal and send it through the machine. An adjustable offset roller can be brought up or down in close proximity to the parallel rollers with the result that a symmetrical cylinder is effortlessly formed. A cone may be formed by tightening up one side of the rollers more than the other.

This is my favorite English wheel. It has casters for portability, it has a built-in rack for anvil wheels, it's as stable as an office building, and the top wheel is supported on both sides for maximum rigidity. If I could hang a hammock across its loving frame, I'd sleep there. Things you want: brakes on your casters, a large kickwheel at the bottom, and a sturdy anvil yoke.

## English Wheel

The English wheel can be used to stretch sheet metal into specific shapes; it can smooth bumpy sheet metal that has been roughly shaped by hand, and it can bend metal into crisp beads or gentle arcs. As you roll a panel between the large flat upper wheel and the lower anvil wheel under pressure, the squeezed panel conforms to the shape of the contact patch of the lower wheel, and you end up trading metal thickness for additional surface area, which manifests itself as additional crown in the panel. The anvil wheels are ground on different radii to allow the curvature of the work piece to increase.

A slip roll is useful for creating cylinders and cones. The metal is passed between two parallel rollers and an offset roller. By adjusting the pressure of the parallel rollers and changing the location of the third roller, the shape of the finished piece can be changed.

We made the exhaust for our Model T speedster with the slip roll. The conical shapes look racier, in my opinion, than straight pipes would have.

I realize that the foregoing list of tools seems like a lot of equipment when taken as a whole, but notice that I described very few items as essential. A sampling of the right tools will allow you to work more quickly and efficiently, but they won't limit your potential to develop as a metal shaper and car restorer. About once every two or three years, I'll get a student who is a natural-born craftsman. These students seem to have developed outside the mainstream, and they come from all over the country. They are not social misfits, but they definitely have a passion and intensity about them that sets them apart from their peers once work has begun. Their tools are inevitably a hodgepodge of random things that they have accumulated, modified, and improved. Although we have a well-equipped shop, these students lug their tools around with them in old toolboxes that they carry on one shoulder and insist on using in class. Predictably, everything these kids touch turns out beautifully—even tools they make for a single use. These über-students are a joy to teach, of course, but they always give me a welcome reality check. The ability to craft things from metal truly boils down to passion, effort, creative problem solving, an ever-growing understanding of how metal behaves, and the desire never to stop learning.

## HOW TO MAKE A SHOT BAG

I will finish this chapter with a section on making a shot bag. I've never seen this procedure described anywhere, so I thought I would include instructions on how we make our shot bags in case some readers feel motivated to do so. I often tell students that if you have a stump, a hammer, and shot bag, you can make just about anything. The shot bag is great for forming and straightening sheet metal, and it can help hold down awkward pieces on the workbench while you work on them. Shot bags are available new from various sources, and they are not particularly expensive. Nevertheless, many metal-crafting folk take great pleasure in making things themselves, so here are some tips.

At our college, we have bags we have purchased and bags we have made. The bag illustrated in the chapter on beginning forming was purchased, and it is the sturdiest design I've come across so far, thanks to an extra layer of support around the top seam. The bag illustrated here is easier to make, however, and will hold up well if you can find a reasonably sturdy piece of leather to use.

One of the benefits of making a bag is the freedom to make several in various sizes. Technically, as long as the diameter of the bag is larger than the face of the largest hammerhead you plan to use, the size of your bag is adequate. Regardless, you will be tempted, and I wholeheartedly encourage you, to experiment with a few different sizes. Once you have them, you will wonder how you ever lived without them. I think about 9 inches in diameter is optimum for your first bag, and if you have some scraps left over, make a small bag that you can hold in one hand. You will love it when you are working with large sheet metal pieces on a workbench or if the panels are still attached to a car. About 1½ inches is a good thickness. Avoid going too large. Imagine trying to lug a pillowcase filled with lead around your shop.

The only trick to sewing a bag is that you usually need a sewing machine with a compound walking foot or you must sew it by hand. In case you are unfamiliar with this terminology, the compound walking foot grips the material from the top and bottom and moves it forward incrementally with each stitch. On most household machines, the material is only gripped from the top side. With thick leather, therefore, the top piece will move forward, but not the bottom, so the stitches will become a mess. In a perfect world, you might gain access to an industrial machine such as the Consew illustrated here. If so, beware. It is tuned for optimum sweatshop use and will sew so fast that the thread practically smokes. Swap out one of the pulleys for a smaller one to slow the machine down so you can control it. In the event you need to crank out a few hundred polyester jumpsuits one afternoon, you can always swap the pulleys back.

If you don't have access to a recently made industrial machine, you might find a usable older machine. The old Singer pictured here may be a little long in the tooth, but I'll bet the old gal has another few decades of productive life left. Plus, the Singer has a distinctive aroma that newer machines will never have. Newer machines are capable of reversing, which is helpful for locking the stitch at the end of a seam. Supposedly holding the material in place and doing a few quick stitches in place accomplishes the same thing when

using an older machine, but it's probably a good idea to tie off the threads at the end of each seam just to be sure.

To get started, cut two identical circles out of a piece of thick leather. Use chalk to mark the leather. Ballpoint pen will almost always eventually bleed through, if you care about such things. Then cut a 2-inch-wide strip long enough to go around your bag with about 1 inch to spare (circumference = 2 × 3.14r); this strip will be the side of the bag. Cut a small piece of stick-on Velcro and sew it in place on one end of the long strip of leather. Sew all the way around the perimeter of the strip of Velcro. Do not be tempted to rely on the adhesive. Lay the perimeter strip along the edge of one of your leather circles. Leaving ⅜-inch for a seam allowance, slowly stitch the perimeter strip to the first circle. Stop sewing about 1 inch before you complete the circle so you will be able to manipulate the loose end. Stick a corresponding piece of Velcro to the strip you sewed a few moments ago. Doing this will prevent you from sewing two identical and therefore nonsticking strips in place, and it will ensure correct alignment. Fold the loose end over and press it into place. You may need to cut off any extra length, as the side strip will often stretch as you make your way around. Pull the Velcro strips apart, and sew the second Velcro strip in place just as you did the first. Close the loose end over again and sew across it. Continue sewing another seam all the way around the bag a second time adjacent to the first seam and tie off the thread.

To finish the bag, align the second leather circle with the edge of the bag side panel. Using a ⅜-inch seam allowance, sew two adjacent seams to affix the top to the side of the bag. Velcro may seem suspect at first glance, but it seals superbly as long as you sew the strips in place. Now fill the bag with sand or No. 9 lead shot. I much prefer shot to sand, but I'm sure other people are just as adamant about sand. Sand is substantially less expensive, comparatively light weight, and readily obtainable. Sporting goods stores, shooting ranges, or gun stores that sell reloading supplies are good sources for shot.

This is an imported Consew industrial sewing machine with a compound walking foot.

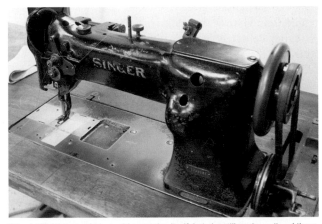

This old Singer is probably as old as sewing itself, but she still works well and there isn't a plastic part in sight.

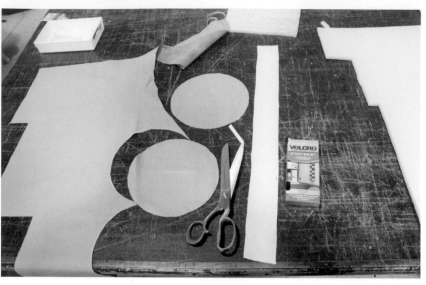

For a shot bag, you need two circles of heavy leather, a strip of leather to go around the bag's side, and a strip of Velcro to seal the bag's opening.

First sew one strip of Velcro onto one end of the leather strip that will make up the bag's side panel. Sew all the way around the Velcro strip to hold it on.

Sew around the edge of the side panel to affix it to the first leather circle, the bottom, but stop short about 2 inches from the end. Find out where the second Velcro strip needs to be, and sew it on.

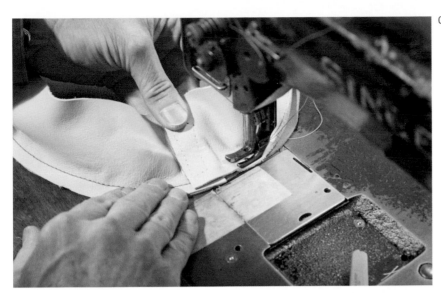

Clasp the two Velcro strips together and sew across them.

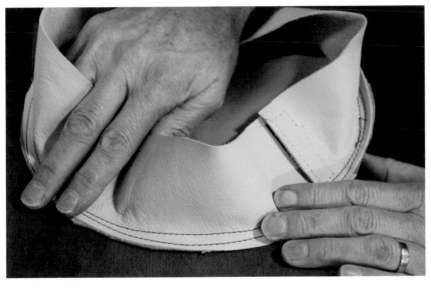

Sew another lap around the bottom panel adjacent to the first seam, and tie off the thread.

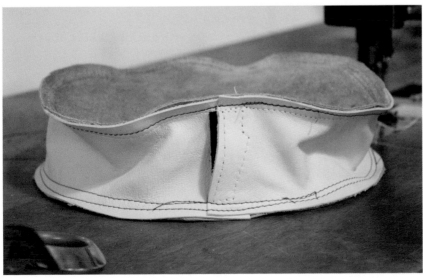

Line up the second leather circle, the top, with the bag side panel, and sew two adjacent seams. Fill the bag with No. 9 lead shot, and you are ready to start hammering.

# Chapter 2
# Special Techniques for Welding Sheet Metal

**B**ecause of its thinness, sheet metal can be difficult to weld—too little heat results in an incomplete weld, whereas too much heat burns a hole in the metal with alarming speed. In addition, sheet metal moves as it is heated and as it cools, thereby creating stresses and strains that can further frustrate your fabrication efforts. With a little guidance and a lot of practice, however, you will start to understand how the metal behaves and how to achieve the results you want. This chapter covers the three welding processes most commonly used for joining sheet metal: the oxyacetylene torch, MIG, and TIG welding. Although this chapter is not a substitute for more comprehensive welding books, such as Motorbooks'

excellent *How to Weld*, by Todd Bridigum, it does cover the use of the torch, MIG, and TIG to solve the special problems inherent to welding thin metals. I encourage readers to follow along in the order that the material is presented. Students new to welding will find that each new technique builds on previous techniques. I hope that readers with prior welding experience will rekindle an interest in, or appreciation for, techniques that they may have fallen out of the habit of using. MIG, TIG, and torch-welding all have definite strengths and weaknesses. Getting thoroughly acquainted, or reacquainted, with them, as the case may be, is a lifelong process that can lead you to new levels of satisfaction and metal crafting merriment.

## OXYACETYLENE OR GAS WELDING

Chances are, if you have ever taken a welding class or looked through a book on welding, you have noticed that oxyacetylene welding is almost universally the first welding process taught to new students. This is true for several reasons: first, mastering the oxyacetylene torch builds manual skills and welding knowledge helpful for learning other welding processes; second, the torch is extremely versatile—some readers may never wish for anything else; and third, the torch is hard to beat in terms of value for your money. There is a dark side to the torch, however, because you are also getting a tremendously dangerous instrument if it is used improperly. Therefore, treat it like you would your neighbor's pit bull, your in-laws, or your local mob boss—with the utmost respect. Fortunately, the welding industry has developed many safeguards to protect users of their equipment, which I will discuss momentarily, but the responsibility for practicing shop safety is *yours*. Review the following precautions until they become second nature to you. For sheet metal welding, the torch is probably the most difficult technique to master, but the finished torch-weld is soft and easy to form, and it behaves like the rest of the panel.

Gas welding uses a mixture of oxygen and a fuel gas to create a flame hot enough to fuse metals by melting them together. Although many fuel gases are used with the torch for a variety of purposes, we will discuss acetylene because it produces a lot of heat, it has a wide variety of uses, and it is readily available. For welding to take place, compressed oxygen and acetylene are supplied from two specially prepared cylinders, each dedicated to one type of gas. The gas in each cylinder is controlled by a dedicated regulator, and

Adam Banks fashioned this Ford Model T steel radiator surround from several hand-shaped pieces that he TIG welded together. TIG was a good choice because it doesn't heat up the work excessively and the welds can be finished to the point that they are imperceivable.

each regulator has two gauges, one to indicate the pressure in the cylinder and the other to indicate the pressure of the gas being fed through a color-coded hose to the torch, typically red for acetylene and green or black for oxygen in the United States. As the gases exit the torch, they are ignited by the user; the resulting flame is then adjusted to the desired intensity. The user then plays the torch flame over the metal work pieces, either with or without a metal filler rod, and executes the weld.

Before you begin torch-welding, you should become familiar with the equipment and its proper use. As a first step, I recommend that you visit your local welding gas supplier, because gas availability will determine your torch-welding options. Find out whether the gas supplier will want you to lease cylinders, buy them outright, or buy the first set and then exchange them as needed after that. The supplier may also recommend acceptable cylinders obtained from another source. The key is that the gas supplier remains happy and willing to fill your tanks over time. For this reason, do not fool around with gas cylinders from unknown third parties, regardless of how inexpensive they are. Cylinders that no one will fill are of little use, and you shouldn't attempt to convince your spouse that they enhance your home décor. Unlike the engine block coffee table and drag slick floor lamp you already own, old compressed gas cylinders threaten more than the standards of good taste—they could leak out all kinds of flammable, poisonous, and otherwise harmful fumes.

I mentioned earlier that several safety features are built in to welding equipment for your protection. Knowing about these features will help keep you safe and will inform your equipment selection. Cylinders intended for welding gases are built according to strict specifications established by the Department of Transportation. These cylinders undergo rigorous testing during manufacture and periodically during their working life to keep them safe. For example, in hydrostatic testing, cylinders are charged with water at pressures exceeding their required limits to ensure that they perform at their rated maximum when filled with compressed gas. The hydrostatic test dates are stamped on the shoulder high on the tank. On older tanks, you will see test dates every five years. Nowadays you will often see a five-point star next to a test date, signifying that the cylinder will need retesting 10 years from the star date. If a cylinder becomes due for testing while in your possession, it need not be taken out of service, but it will have to be retested before it is refilled. If you obtained the cylinder from your local gas supplier in the first place, this will not be problem.

Oxygen and acetylene tanks differ in their construction because of the requirements of the individual gases, but both have rugged threaded steel caps covering the cylinder valves at the top of the cylinders. Because the gas in these cylinders is under pressure, the cylinder valve could be broken off if a tank were to fall over, thereby causing the tank to launch across your shop like a missile. Keep the cylinders upright with the caps in place whenever the tanks are moved or a regulator is not attached, and keep the cylinders chained in position whether they are on a welding cart or standing against the wall. Never allow the cylinders to stand around your shop unsupported, regardless of whether they are full or empty. The caps are not necessarily interchangeable from one tank to the next because of the diameter of the threaded portion of the cylinder and the thread pitch varies; it may be either fine or coarse. I find it helpful to provide a mount for the caps on each welding cart so that they are always handy. Also, if you have multiple tanks, clearly mark the empties to avoid the aggravation of swapping out empty tanks for other empty tanks.

The hydrostatic test dates stamped into this oxygen cylinder indicate that it was tested previously in December 1999 and again in January 2010. The star indicates that it will not need additional testing until 2020.

Oxygen tanks and the cylinder valves that accompany them are different from their acetylene-specific counterparts for two reasons: 1) oxygen is stored at extremely high pressure—2,200 pounds per square inch of tank area, whereas acetylene is stored at 250 pounds per square inch, and 2) acetylene is unstable when stored in a gaseous state at pressures above 15 pounds per square inch. Because of the high tank pressure, the cylinder valve on an oxygen cylinder is a back-seating valve, which means that when it is opened all the way, an inner seal prevents oxygen from leaking out around the valve stem at the top of the valve. Thus, you should always open the valve all the way on an oxygen cylinder to engage this inner seal. Even though the cylinder will flow fully at a full turn or so, operating the valve at this position can allow oxygen to be forced out around the valve stem beneath the hand wheel. Furthermore, an oxygen cylinder valve contains a pressure relief device in the form of a thin metal disc that will give way in the event the tank pressure reaches the point where the tank might rupture. If the disc yields, oxygen escapes from the cylinder through small holes in the disc housing that protrudes from the side of the cylinder valve opposite the normal outlet. Obviously, the cylinder would probably buck and spin violently about, spewing oxygen, in the event the disc were called into service, but this is preferable to an exploding tank or the rocket that would result if the entire cylinder valve came off. Because oxygen violently promotes combustion, keep grease, oil, and other flammables away from your equipment, your clothing, and your work area. In the event you inadvertently open an oxygen valve and oxygen soaks into your clothing, change your clothes and hang up the contaminated clothing for several hours to allow the oxygen to dissipate. Oxygen cylinders range in sizes from 20 to 300 cubic feet. Discuss your anticipated needs with your welding gas supplier, and he or she will be able to advise you which cylinder size would work best for your application.

Although acetylene is stored at a much lower pressure than is oxygen—250 pounds per square inch—it has its own peculiar hazards. Acetylene is unstable when stored in a gaseous state at pressures above 15 pounds per square inch. Therefore, *never* adjust the acetylene pressure on your welding, heating, or cutting torches above 15 psi! To store acetylene, cylinders must be filled with a porous filler and charged with acetone, which absorbs the acetylene in solution until it is withdrawn. In the event an acetylene tank is subjected to high temperatures, the internal pressure will rise dramatically. As a safeguard, the top and bottom of acetylene cylinders have threaded brass fuse plugs with centers made of an alloy that melts at 212 degrees Fahrenheit, thereby releasing the cylinder's contents. A sudden release of acetylene is a frightening thought, but probably a better alternative to an exploding cylinder and the accompanying shrapnel.

Unlike the valve on the oxygen cylinder that must be opened completely, the valve on an acetylene cylinder is opened only about a half turn. As in the case of oxygen cylinders, keep acetylene cylinders upright and chained in place at all times. Because of the acetone inside, leaving an acetylene tank on its side can concentrate the acetone near

This is the proper way to store compressed gas cylinders—chained to the wall with the steel cylinder caps in place. Note that one cylinder wears a metal tag to indicate that it is empty.

the top of the cylinder, which can be detrimental to the regulator if the torch is used right away, not to mention you will be consuming the acetone that is there to stabilize the cylinder. If you must right an overturned tank, let it sit for several hours before use so that the acetone can recede back into the cylinder. To keep the acetone in the cylinder where it belongs, you will find that you need to watch the tank pressure gauge more closely with acetylene than with oxygen. Whereas a direct correlation exists between the cylinder pressure and the amount of oxygen left in an oxygen tank, the acetylene cylinder pressure gauge remains constant for a long time and then drops more quickly. This action is because the acetylene leaves the cylinder at a steady rate until

it is almost gone. The amount of acetylene in a cylinder is determined by weighing the cylinder and comparing that weight with the tare weight, the weight of the cylinder without acetylene. The tare weight is stamped on the cylinder. In its rightful place, acetone stabilizes the acetylene in the cylinder, but once it exits the tank it can deteriorate rubber parts in the regulator or torch, it cools your welding flame, and it degrades weld quality. If your flame looks purple or the torch literally drips fire, you are burning acetone. Go ahead and refill your cylinder once the needle on the high-pressure gauge is low. If you are paying attention when you first turn on the acetylene cylinder valve at the start of each welding session, you are not likely to run the tank too low.

The small protuberance on the side of this oxygen cylinder valve houses a built-in safeguard, a disc that will rupture in the event of high temperatures would otherwise rapidly elevate cylinder pressure. As another safety precaution, the back-seating valve described in the text is housed beneath this hand wheel.

In this view of an acetylene cylinder, you can see one of two fuse plugs located at the top of the cylinder designed to bleed off the cylinder's contents in an emergency. The numbers stamped into the cylinder include its Department of Transportation identification number, its capacity in cubic feet, and its tare weight, the weight of the cylinder without gas.

Acetylene tanks come in various sizes from 10 to 380 cubic feet. Portability and the frequency of refilling are factors to consider when choosing cylinders, but the flow requirements of certain welding, heating, and cutting tips are even more important considerations. Because acetylene can be consumed only as quickly as it can be released from the acetone solution inside the cylinder, equipment requiring high flow rates can easily overtax the available acetylene in the smaller tanks and thus draw acetone out of the cylinder. The general rule is that the maximum safe flow rate of acetylene is ⅐ of the cylinder's capacity per hour. Therefore, a tiny 10-cubic-foot cylinder can safely supply only about 1.4 cubic feet per hour (10 cubic feet divided by 7). A welding tip for ⅛-inch steel, for example, needs acetylene at a flow rate of about 9 cubic feet per hour. Thus, a theoretical 63-cubic-foot cylinder would be the minimum safe choice. The only time this will be a problem with automotive applications, however, is with the tiny portable welding kits. Charts showing the relationship between metal thickness, torch tip size, and gas flow requirements are commonly available—a chart like this will certainly be included in the manual that comes with any new torch outfit. If you plan to weld, cut, or heat thicker steels, peruse the charts and then consult your gas supplier for the appropriate-size cylinder.

Transport your tanks upright and secured in the back of a truck, *not* in the trunk or passenger compartment of a car. A small leak could displace the oxygen in the car or contribute to a catastrophic fire. A safety poster I have seen in many welding supply stores shows a car that has had a compressed gas cylinder explode in its trunk. I hope you will have the opportunity to see it as well; you will not be tempted to risk repeating this disaster.

Basic torch outfits are widely available and are not very expensive considering their versatility and long life. Although good-quality welding equipment is rebuildable, I would not buy used equipment as a beginner unless you are buying from a welding supply store or you are familiar with the equipment's history. Too many potential gremlins could rear their ugly heads, and you may not realize you have a problem until you have wasted a lot of time and become frustrated. Furthermore, when you shop for a torch outfit, consider whether additional torch tips, replacement parts, and service are available in your area. Most welding supply stores carry the products of well-established companies, so you will be able to keep your equipment in service. High-quality equipment, well treated, will last a very long time.

## SETTING UP YOUR EQUIPMENT

Leave the cylinder caps on until both the oxygen and acetylene cylinders are chained in place on a welding cart or against a wall if you will do all your welding from one station. Unscrew the caps and turn each cylinder so the valve is in the most advantageous position for viewing the regulators once they are installed. Stand to the side and crack each valve slightly to

clear out any debris that may have settled on the valve seat. Wondering if this was really necessary, one time I did not do this. I attached the regulator and found that I had a small leak. I removed the regulator and found a tiny speck of dirt on the seat that had caused the leak. Once I blew it off, all was well. If you examine the regulators, you will notice that the acetylene regulator has a groove running through the flats on the nut that secures the regulator to the cylinder valve. These same grooved nuts may be found on the acetylene hose fittings. The grooves signify that these fittings are *left-hand thread*, so they tighten to the left, which is opposite of just about every other threaded fitting you encounter. This simple but effective safeguard prevents mixing gases. In addition, the nuts attaching the regulators to the cylinder valves are different sizes for oxygen versus acetylene. In addition, the former is a female fitting, whereas the latter is male. Tighten the regulators on the cylinders with a wrench nice and firm, but not so tight that you risk damaging the soft brass threads. A cylinder wrench is ideal for this purpose, a fixed-end wrench is good if you have one large enough, and an adjustable wrench is satisfactory as long as you apply torque toward the adjustable jaw to avoid damaging the nut. Do the same with the hoses and torch handle. If you are working in the United States, remember to use the red hose with *left-hand threads* for acetylene and the green or black hose with *right-hand threads* for oxygen. The torch handle will be stamped to indicate which hose goes into which inlet, usually *OXY* for oxygen and *GAS* or *FUEL* for the acetylene in our case.

Find the appropriate-sized welding tip for the metal you want to weld by looking it up in a chart. Tighten the welding tip on the torch handle hand tight only! This practice is to protect the rubber O-rings that seal the tip to the handle. Make sure the pressure screw on each regulator is loose. You can tell as you start to screw it in when it engages the diaphragm because you'll feel spring pressure against it. Then screw each regulator onto its cylinder by hand. To avoid cross-threading the regulator nut on the cylinder valve, jiggle the nut from side to side as you screw it on. Make absolutely certain that the regulator nut engages the threads smoothly before you apply any torque with a wrench. Once the regulators are tight, double-check the pressure screw on each regulator to make sure it is loose. Having the pressure screw tightened down when cylinder valve is opened shortens the life of the regulator because the full tank pressure slams against the diaphragm. Although a regulator bursting apart is a highly unlikely occurrence, stand to the side when you open cylinder valves. Slowly open the acetylene cylinder valve about a quarter to a half turn. Doing this should provide the necessary flow but still be easy to turn off in the event of an emergency. When I began welding, the acetylene cylinders I used were opened with a small ⅜-inch square wrench rather than a hand wheel. At the time I hated these in comparison to a hand wheel. The wrench seemed cumbersome by comparison, and square

To prevent the unintentional mixing of oxygen and acetylene, all regulator and hose fittings for acetylene are left-hand thread (tighten to the left), whereas the oxygen fittings are right-hand thread. The groove machined into the fittings is a subtle reminder of this fact. In addition, the torch handle inlets are marked *OXY* and *GAS*.

From left to right: open-end wrench, adjustable wrench, and cylinder wrench. Adjustable wrenches are satisfactory for use on gas fittings as long as you remember to apply the torque in the direction of the adjustable jaw to avoid rounding off the soft shoulders. Naturally, a cylinder wrench is like a self-contained toolbox for gas welding.

fittings on the tanks were often worn. Thankfully, I have not had to use one of these tanks in years, but if they are used in your area, leave the wrench in place atop the cylinder valve for quick access. Gently open the oxygen cylinder valve and keep turning it until it is all the way open—remember the back-seating valve? Consult the same chart you used to determine which welding tip to install and find the recommended gas pressures for that tip. Tighten the pressure screws on the low-pressure gauge on each regulator to set the recommended pressure. Open the valve for each line one at a time to set the pressure of the torch as if it were in use. The static pressure is always a little lower than the actual working or dynamic pressure. The variance grows as the flow demands of the tip increase. Notice the peculiar garlic-like smell when you crack the acetylene open; remember it so that you will know if ever you have an acetylene leak in your shop.

When hooking up a torch for the first time, and every time you swap out cylinders, leak-test all your regulator and hose fittings by spritzing or brushing them with soapy water. Leaks will begin to bubble. You must correct the problem before proceeding. You may need to go back and use Teflon tape or pipe thread sealant on pipe-threaded fittings. If the

After determining that the pressure screw is backed off on the regulator, always stand to the side when opening any high-pressure cylinder valve to protect yourself from debris if the regulator were to come apart. In addition, open the valve gently to prevent suddenly applying the full tank pressure to the regulator diaphragm.

regulator leaks at the cylinder valve, the valve is probably damaged. Do not harm your regulator by overtightening it in an attempt to compensate for a bad cylinder valve. Take back the cylinder and exchange it for another. As you are leak-testing your equipment, glance at the low-pressure gauges on the regulators to make sure the pressure is not creeping up. If it is, an internal leak is in the regulator. Realizing this would be particularly important with an acetylene regulator because the pressure of the gaseous acetylene could rise to dangerous levels very quickly. Have the defective regulator repaired immediately. Once you have successfully leak-tested your equipment, you will have also purged the gas lines by running the respective gases through each hose one at a time, thereby ensuring that there is no mixture of gases in either line. Any time you begin a new welding session, it is a good idea to purge the lines as well.

Every time you change out a cylinder, make any substitutions to your equipment, or long for some peace of mind, apply soapy water to your gas fittings. Leaks will bubble noticeably.

## LIGHTING AND ADJUSTING THE TORCH

To light the torch, make sure the gas pressures are correct as described in the previous steps, aim the torch in a safe direction, crack the acetylene valve at the base of the torch slightly, and ignite the acetylene with a striker. With low acetylene pressure at the tip, you will end up with a large soft orange flame that generates soot like a burning sofa. Increase the acetylene by opening the acetylene valve at the base of the torch handle until the black soot dissipates. The end of the flame will be turbulent, but the flame will still taper back into the torch tip. This illustrates the correct amount of acetylene. If you keep adding acetylene, the flame will get louder and an air gap will form between the end of the welding tip and the beginning of the flame. If it looks as if the flame starts about ¼-inch out from the tip, you have too much acetylene pressure; the flame will extinguish itself as soon as you try to add oxygen. When dealing with thin sheet metal, you may find that one size tip seems a little too hot and the next-smallest size tip does not seem hot enough. In that case, turn down the acetylene pressure slightly on the larger tip. The flame may not be as turbulent as the flame you see in every picture of the correct flame, but you should be able to get the flame you need. There is a lot of heat control to be had by varying the intensity of the pure acetylene flame before adding oxygen.

When you feel you have established the correct amount of acetylene you may start adding oxygen by opening the oxygen valve at the base of the torch handle. You will notice a color change immediately. The flame will turn predominantly blue, and three distinct zones will appear. The outer, amorphous part of the flame is the envelope. A secondary zone, called the feather, will form inside of the envelope. This kind of flame is called a carburizing flame because it contains more acetylene than oxygen and tends to add carbon to metal that is welded with it, thereby producing brittle welds. As you continue to add oxygen to the carburizing flame, the feather retreats toward the torch and a small cone will become obvious right near the torch tip. As soon as the feather recedes and meets the cone, you have achieved what is called a neutral flame,

which is ideal for welding. If you continue to add oxygen to a neutral flame, the envelope and cone will get smaller and the torch will hiss angrily. This last flame is known as an oxidizing flame because it contains more oxygen than acetylene.

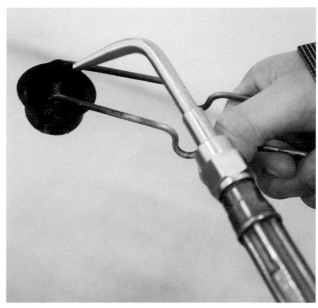

To light the torch, crack the acetylene valve at the base of the torch handle, place the striker alongside the torch tip, and create a spark. Do not use matches, cigarette lighters, or other people's torches. Cupping the striker over the torch tip can produce a dramatic, hair-singeing flare-up.

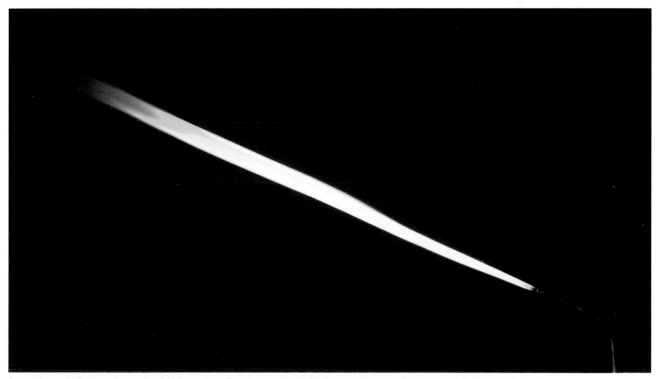

An acetylene-only flame produces gobs of thick soot that will coat your work area in a hurry, so do not spend any more time in this mode than necessary.

To adjust your torch for welding, add acetylene until the soot dissipates. Do not be concerned that it takes a little interpretation on the user's part. Sometimes with thin metals, you will need to adjust your torch down at the lower end of this range.

If you have too much acetylene, the flame will be loud and you will see a gap between the base of the flame and the torch tip.

As you begin to add oxygen, you will create a carburizing flame with three distinct zones: a large soft envelope, an inner feather, and a very small cone just off the tip. This flame is acetylene-rich.

Continue adding oxygen to the carburizing flame until the feather recedes back and joins the inner cone. This is called a reducing flame or neutral flame. It is the flame you will use for welding.

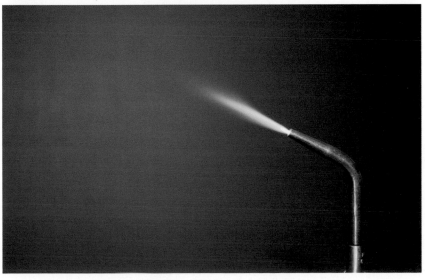

Adding more oxygen to the neutral flame creates an undesirable oxygen-rich oxidizing flame. It will be loud, the envelope will be short, and the cone will be smaller than before.

## SHUTTING DOWN THE TORCH

After you have experimented with adjusting the torch to achieve carburizing, neutral, and oxidizing flames, you may shut off the oxygen valve at the base of the torch handle. Then do the same for the acetylene. The flame at the torch tip should go out. Now use the hand wheel at the top of each cylinder valve to turn off the oxygen and acetylene. Gas is still present in the regulators and hoses, of course, which needs to be bled off for safety's sake and to prolong the life of the regulators; otherwise their internal diaphragms would be perpetually pressurized. Bleed off the pressure in the lines by opening the valves on the torch handle one at a time. As you hear the gases exit, you will see first the high-pressure gauge on each regulator go to zero, followed by the low-pressure gauge. After each set of gauges goes to zero, turn off the valve on the torch and back out the pressure screw on each of the regulators. You have successfully bled off the hoses and released the pressure on the diaphragm inside each regulator. Practicing this procedure will greatly prolong the life of your equipment.

## HOW TO WELD

Obtain some clean scrap steel the same type you plan to use for your sheet metal projects. Even if you plan to use aluminum eventually, I would recommend starting with steel because it is much easier to see what is taking place as you heat the metal. The presence of paint, undercoating, dirt, rust, decals, old body filler, oil, or any other foreign matter will prevent you from making sound welds. Likewise, avoid welding over old leaded repairs, cadmium plating, or galvanized steel because noxious fumes will be given off. The shape and size of the metal pieces will not matter much as long as you have some clean edges to butt together or you have the tools to create some clean edges, like a band saw, belt sander, shear, tin ships, or cut-off tool. You will want the thickness of your metal to approximate the type of sheet metal you envision working with on your future projects. The thickness of the metal you will weld determines which welding tip to use and, consequently, which gas pressures to select. As I always tell students, this kind of information is available on a chart, so there is no excuse for guessing. You can find welding charts in every welding book, in free pamphlets from your welding supply, and, of course, on the Internet. Each manufacturer of welding equipment has its own tip numbering system, so find the chart for your type of torch. Learning to weld on thin sheet metal is tough if this is your first welding experience, but the more time you invest learning to weld your chosen material the better off you will be down the road. Now you need a surface upon which to weld.

While custom-made firebrick welding tables are nice, a homemade metal table will work just as well. Get a flat piece of steel and lay it across a couple of sawhorses. As long as you do not apply massive amounts of heat to it, a piece of thin sheet metal over a sheet of plywood works well too. You will also want some kind of blocks to lift your work piece up off the table. They could be firebricks, pieces of square tubing, or parts of I-beams—anything inflammable that will lie flat and act as a steady base for your work piece. If your piece lies directly on your welding table, the table will act as a heat sink, robbing vital welding heat from your work piece and wasting gas. Expanded metal could be used as a worktable, but things tend to fall through it. Make sure your work area is clean and clear of flammables! Do not be tempted to lay something across a metal trash can for an improvised welding table. Many of us have done this at some point in a moment of metal-crafting delirium. It always results in a fire, the severity of which depends on the volatility of the solvent-soaked rags that are, but shouldn't be, at the bottom of the trash can. This reminiscence reminds me of another admonition: have a fire extinguisher handy, and know how to use it.

Once your work area is clear and safe for welding, you can begin using the torch. For most torch-welding, I wear a face shield with a No. 5 tinted shield because welding goggles do not fit well over my massive safety glasses. No. 5 is a typical all-around shade, but sometimes it is nice to have a little more or less protection depending on the intensity of the light you are dealing with. I like thin TIG welding gloves for everything except the heaviest cutting, in which case I might step up to the insulated oven-mitt welding gloves that you typically see at home improvement stores. Otherwise, I find the insulated gloves too much like oven mitts in their ability to block out all feeling. I also wear a long-sleeve cotton shirt, blue jeans, and leather lace-up boots. Lay a flat piece of steel down on some firebricks or blocks on your table, light the torch, and adjust it to produce a neutral flame.

The manner in which you hold the torch for welding is purely a matter of personal preference. Some people hold it like a giant pencil; others simply grab onto the torch handle like a cane fishing pole. Experiment with different grips as you play the flame over your practice piece. Create a puddle by swooping down with the torch until the inner cone is slightly above the steel, about ⅛ inch. With thin sheet metal, you do not want to spend any more time heating the metal than necessary, so get in the habit of being deliberate. I have noticed that students are often tentative with the torch because they do not want to burn holes. They hover the torch 1 to 2 inches above the work. As a result, the heated portion ends up being much too large for a weld and the metal warps fiendishly. With your torch tip close to the work piece, you will see the steel rapidly change color. You will see several subtle color variations as steel is heated, but the main ones you will notice right away are red, orange, yellow, and white, and then you will burn a hole in your work piece. This entire process will take only about 10 to 12 seconds or less with sheet metal. Years ago at a

Inexpensive shade No. 5 welding goggles, lower left, are a good choice for most torch welding and cutting, but they can be problematic if you wear glasses or are prone to heavy perspiration. Like goggles, full-face shades are available in a range of shades and will last a long time if you remember never to lay them face-down on the lens portion. Lightweight TIG welding gloves, middle, provide adequate protection plus superior feel to the heavier oven mitt gloves at right.

blacksmithing workshop, I was taught that the appropriate time for hammer welding was when sparks started to jump off the steel. I have noticed this phenomenon with the torch as well; right before you burn a hole, the metal will be white and sparks will dance off the surface because you are starting to burn the metal. Perhaps this observation will prevent you from burning too many unwanted holes. Bring the torch in toward the steel again, but this time, as soon as you see a shiny puddle form, move the torch steadily from right to left if you are right handed or the opposite if you are a lefty. Do this until you can create a puddle and push it across the steel without burning through.

If your torch makes a loud popping sound while you're welding, you've just experienced a backfire, or torch pop, which results from gases igniting inside an overheated torch tip. If the tip is too close to the work or you are welding in a confined corner, the tip can get hot enough to ignite the gases prematurely. Carbon buildup in the tip, touching the tip to the work, insufficient gas pressures, or weld spatter flying back up into the tip orifice can cause torch pop. Clean the torch tip, let it cool, and try increasing your pressures

about 1 psi each. If while welding your torch goes out and you hear a shrill whistle, you have just caught a glimpse of the welding phenomenon known as flashback. This flashback has nothing to do with those raucous times you had in the 1960s; it is when the welding gases start to burn inside the torch or hoses. A true flashback can be extremely dangerous because the flame continues to travel up the hose. If you hear a whistle or hissing sound and the torch flame goes out, immediately turn off the oxygen cylinder valve and then the acetylene cylinder valve. Try to determine if the torch simply touched the work—sometimes you'll get this mini-flashback instead of torch pop—or whether the tip got too hot. The torch might be starved as a result of maladjustment, or you could have a kinked or clogged hose. Having been warned about the potential problems you might run into, you are now ready to run a weld bead.

To run a bead across your work piece, bring the torch down until the inner cone is approximately $\frac{1}{16}$ to $\frac{1}{8}$ inch above metal. Tilt the torch tip at about a 45 degree angle with the tip leading the way. Too much angle doesn't heat the metal enough; too little angle will tend to create turbulence in the

Some welders hold the torch like a pencil, draping the hoses over their arms.

I call this the *"goin' fishin" grip* because it reminds me of sunny summer days with a cane pole in hand. Feel free to improvise your own grip. As long as you have a safe hold on the torch and can move it comfortably, you will do your best welding.

Create a pool with the torch by bringing it down to the work piece at a 45 degree angle until the inner cone is about ⅛ inch above the metal. A molten puddle should form within seconds of the torch being hot enough. Practice moving this puddle along the metal in a forehand direction (toward you).

weld pool and will be more likely to burn a hole. Keep this in mind so that you can vary the angle a little on either side of 45 degrees to add or decrease heat. Moving the weld pool forward ahead of the tip is called forehand welding. Backhand welding is when the torch leads and the weld pool follows behind. I have read that it is recommended for welding thin metal, but I have never liked it as well as forehand welding. On thin metal, I like the psychological closure that comes with leaving the weld and the super-heated metal behind. Also, the torch does not block your view of the seam as it does in backhand welding.

The time-honored technique for manipulating the torch is to move the torch tip in a tight circular pattern as you go, keeping the cone within the borders of the weld pool. With thin sheet metal, however, I'm not convinced that swirling is necessary. It seems to me that once you get both sides of a joint hot enough, they tend to flow together naturally. As long as you get complete penetration, it is a sound weld. Experiment with swirling and not swirling as you move the puddle around on the surface of your panel to see what you think. As I often tell students, there is more than one way to skin a cat. I don't know and don't want to know the origin of that phrase, but I am convinced that there are many possible solutions to each problem. As long as safety is not compromised, I encourage experimentation. Sometimes the best way to do something is contrary to the standard theory. Practice running several beads on a single panel before attempting to join two pieces of metal.

This torch weld was produced by swirling the torch in a circular fashion. Do not be concerned if your sheet metal torch welds resemble the *travel speed too fast* illustration you remember from your high school metal shop text. Automotive sheet metal is thin, so do not be surprised to find that you need to move the torch along briskly to keep from burning holes.

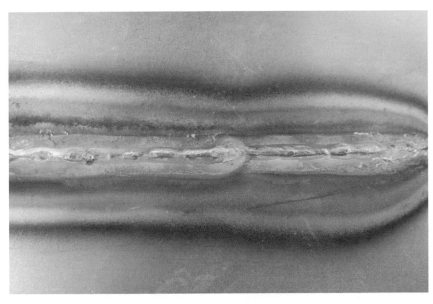

Examine the back side of the weld to see if you have achieved penetration. If in doubt, clamp your sample in a vice and hammer it over on itself from the back side to test the weld.

## A FLANGE WELD WITHOUT FILLER ROD

A flange weld is a good introduction to joining multiple panels because it does not require the use of a welding rod. Bend a 90 degree flange in two pieces of sheet steel and lay the pieces on your worktable with the flanges back to back. This setup will allow you to fuse the edges of the two panels using only heat. Utilize Vise-Grips or clamps to keep the flanges tight against one another. Use a torch with a neutral flame to make several small tack welds along the length of the seam to maintain alignment. Employing the same technique you used earlier to run some beads on a flat panel, create a puddle at one end of your seam and carry it all the way down the length of the seam. As you approach the end of the seam, you may find that laying the torch down almost parallel with the seam will help you keep from burning a hole at the end. Do

not be concerned with burning holes at this time. If you burn a hole, simply remove the torch, let the metal cool some, and continue welding ahead of the hole. Filling holes is a job for welding rod. Practice the flange weld until you can weld the entire seam without burning holes.

## FILLER RODS

Welding rod is added to a weld whenever additional metal is needed along a weld seam to ensure that the metal along the weld is at least as thick as the parent metal. When you butt two pieces of metal together and fuse the gap, you remove metal from the edges of the two original pieces to fill the joint. Clearly, the metal along the seam will not be a uniform thickness in this case. A small concave bead is not necessarily a problem, even with thin sheet metal, however, because often

To weld without a rod, bend a 90 degree flange into two separate scrap pieces and weld them together along the flange. The clamps will help hold the metal until you have established several tack welds.

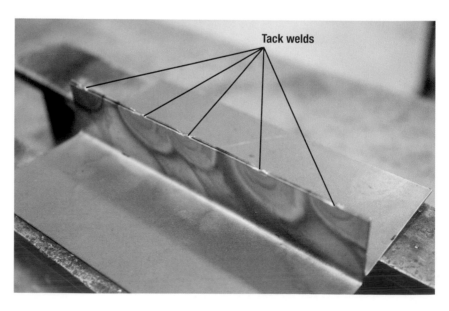

After establishing good alignment, the next step in any weld is to put down a few tack welds to keep the metal in position.

Tack welds

If you are right-handed, start your weld at the first tack on the right end and move toward the opposite end, and then finish the seam where you started. Any time you are welding near the edge, you may find that it helps to lay the torch down almost parallel with the seam to prevent applying heat where you don't want it. Remember, the torch is only a tool for applying heat to do the work you want. Do not be afraid to experiment.

you will undertake additional steps of surface finishing to prepare the metal for paint. If, on the other hand, the metal must be finished to a level where no evidence of welding can be seen, then enough metal must be added to build up the joint to the point that you can file it down smooth for a seamless, invisible repair.

Welding rods come in a wide variety of diameters and grades and are either prepackaged, or can be purchased by the pound from your welding supplier. If you are a beginner, do not buy any better rods than you need while you are learning. If you weld race car or aircraft parts, get the best rods that

you can for your application. Welding rods are manufactured of specific alloy combinations for various purposes. Do your research to make an informed decision. The American Welding Society classifies welding rods according to their composition and tensile strength. Classification numbers are usually stamped onto the larger rods. For small-diameter rods, you will have to consult the package label to identify the rods. For RG 45, for example, the R stands for welding *rod* and the G stands for *gas* welding. The 45 denotes the approximate tensile strength of the weld in thousands of pounds per square inch—45,000 psi in this case. The rods

Many filler rods have their American Welding Society designation stamped into them for identification. From top to bottom, here are examples of a gas rod, two aluminum rods, and a stainless rod. Small-diameter rods are often not marked, so come up with a storage solution that keeps them organized and out of the weather.

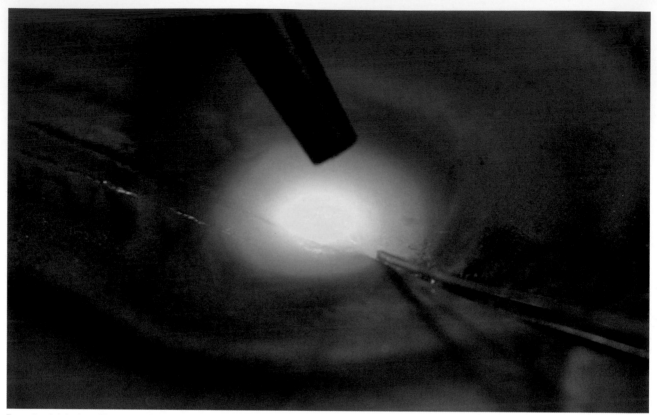

To create a bead using a filler rod, create a puddle and bring the rod down in close proximity to the pool, but not so close that it will melt. As you advance the puddle, dab the rod in the leading edge of the puddle. Do not fixate on the rod, but concentrate instead on the weld pool.

are coated with copper to minimize wear on the tooling used in their manufacture. Copper may help with rust prevention as well, although copper-coated rods will develop surface rust over time if they are handled and left sitting in the open air.

Begin practicing with a welding rod on a flat piece of steel before trying to weld a joint. Select a rod with a diameter that closely matches the thickness of the metal. Bring the torch down, make a puddle as before, and start moving it across the work piece. Bring the rod in to within ½ inch of the torch tip but do not touch the torch tip. The trick is to keep the rod close enough to the tip that it will be preheated and therefore ready to melt as soon as it is dipped in the weld puddle, but not so close that the rod melts prematurely. If the rod melts while it is hovering above the weld pool, it will deposit globs of metal on the surface of the parent metal. The dripping metal from the rod will be cooler than the parent metal and will not fuse properly to the surface. If the rod is held too far from the weld zone, the rod will cool off. Then when you try to add metal to the weld puddle, the cool rod absorbs heat from the puddle, consequently cooling the puddle and probably sticking in the puddle. If the rod sticks, heat the end to release it. When you can successfully run a few beads using the rod, attempt a butt joint, where two pieces are butted together and welded along their edges.

## BUTT JOINTS

The butt joint is commonly used in fabrication because it can be finished to a high degree and does not have corrosion-prone overlapping sections. Lay two pieces of clean scrap sheet metal on firebricks or blocks on top of a welding table. If you begin with the edges of the pieces to be welded literally butted together, the pieces will inevitably expand as they are heated and contract as they cool, and one piece will climb over the other or they will warp and buckle. To minimize this effect, start with a wider gap at the opposite end of the seam from where you start. For a 12-inch panel, for example, the gap should be about ¼ inch at the far end of the panel. Butt the panels together at the starting end and clamp them semitight so they can still move. Clamp the opposite end just tight enough to hold the panels in the same plane. Make a series of tacks along the seam about ¾ inch apart. Placing a tack just past the heat zone of the previous tack seems like a magical formula that works really well. If the panels don't puddle together at the tacking points, add a little filler rod right away to establish the tacks without burning through. At any point, if the pieces of metal wander out of the same plane, hammer them back into alignment. You may have to separate some tack welds to maintain alignment. Vise-Grips are not a must, but can be very helpful for keeping the pieces of metal aligned.

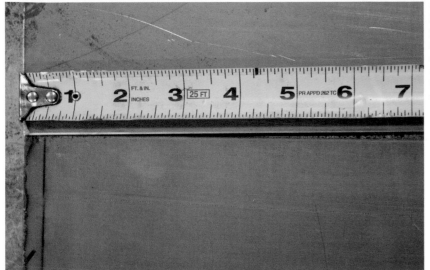

Getting the gap just right for torch welding will take a little practice. Butt the panels together at the end where you start your tacks, but leave a gap at the opposite end of the panels. A ¼-inch gap at the end of a 12-inch panel or ⅛-inch gap for a 6-inch panel seems just right. Too much gap will lead to burning holes. Too little gap results in the metal forcefully butting together, warping, and/ or overlapping.

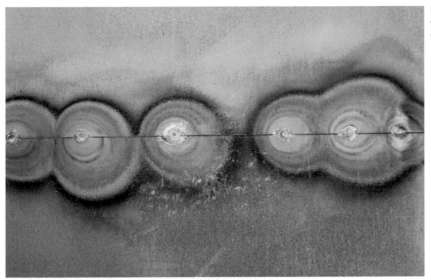

Notice how the gap has largely closed after tacking. The tack welds will help hold the panel in alignment, and they will help dissipate the welding heat out from the seam.

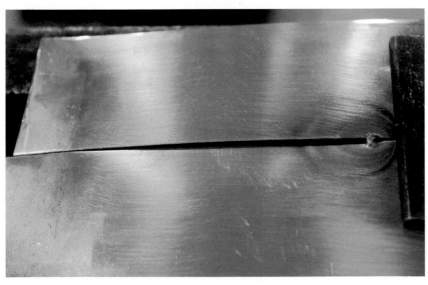

As you apply heat, metal tends to wander out of alignment. Correct it immediately.

41

I caution students to stay in control of their welds throughout the welding process. By that I mean that you should stop welding the moment any misalignment occurs. You will be tempted to complete welds that are only slightly awry. Trying to finish out metal that has been welded out of alignment is next to impossible, so just stop, correct the misalignment, and retack the seam. Once you establish a series of tack welds about ¾ inch apart along the length of the seam and correct any misalignment issues, you may run a continuous bead along the seam. The tack welds will hold the panels in alignment and act as conduits to transmit the welding heat out from the weld area. Keep in mind that adding the rod means that there will be more metal to grind away.

One might think that welding small areas and allowing them to cool before proceeding would be the best way to put the least amount of heat into the metal. This is exactly how I would approach a MIG or TIG weld, which I will discuss in later sections. However, at least for me, performing a torch-weld in one continuous operation helps me tap into a certain physiological and psychological groove that produces more consistent welds. When I start and stop a dozen times with the torch, allowing the metal to cool each time, I inevitably get more unevenness in my bead and more warping in the metal. The blue heat zone on either side of the weld will show evidence of uneven heating. When I run a continuous bead, however, the heat zone stays nice and even and the metal moves less.

To keep from burning away the metal at the end of the seam, tilt the torch back so that the torch is almost parallel to the work piece, and be prepared to add rod at the last moment. A successful torch-welded butt joint between two pieces of thin sheet metal will inevitably have some unevenness due to heat distortion. Do not be overly concerned at this time. One of the great benefits of a torch-weld is that it is extremely malleable and therefore can be hammered and filed to improve the appearance and shape of the work after welding. I will cover this topic in the chapter on metal straightening. Although there are other types of welded joints that one may perform with a torch, the butt joint is the most widely used for the kinds of projects covered in this book. I refer you again to Todd Bridigum's *How to Weld* for more instruction on general welding. Now, let us apply what we have learned to torch-welding aluminum.

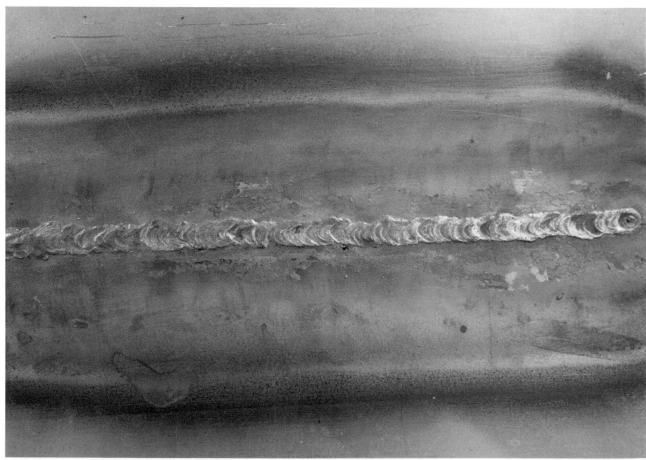

This close-up of a finished bead made with a rod shows how much metal has been deposited along the weld seam.

Amco 22 aluminum flux greatly facilitates torch welding aluminum by preventing oxide formation along the weld zone.

## TORCH-WELDING ALUMINUM

Torch-welding aluminum is very similar to torch-welding steel after you get over the initial shock of overheating your first piece of metal. Unlike steel, aluminum does not change color as it is heated, so it is easy to overheat. Another difference between torch-welding steel and aluminum is that the oxides that form on the surface of aluminum melt at a higher temperature than does the parent metal. Consequently, you must use a flux such as Amco 22 aluminum welding flux to protect the weld from oxide formation. Without it, the surface of the aluminum acts like a rubbery membrane holding the parent metal together, refusing to pool as it is heated until the whole mass collapses in a molten glob. With the right materials and equipment, however, you will need only a little more practice to get accustomed to aluminum. Because an aluminum torch-weld inevitably anneals a large area of metal along the weld seam, torch-welded aluminum panels are easy to shape and finish. Keep in mind, however, that an aluminum TIG weld can be rendered ductile by annealing, if necessary, so there is nothing inherently better about an aluminum torch-weld. The technique is worth learning if the torch is the only equipment at your disposal, but do not let

anyone convince you that the torch is somehow superior to another process.

For some welders, hydrogen is the fuel gas of choice for welding aluminum because the oxyhydrogen flame is over 2,000 degrees Fahrenheit cooler than an oxyacetylene flame. Because aluminum melts at 1,271 degrees Fahrenheit, they reason, why use a 6,300 degree oxyacetylene flame to weld it? This is sound logic, but for some of us, having to obtain one more set of gas cylinders and regulators for hydrogen is just too much hassle, especially when the oxyacetylene torch will work, overkill though it is. If you already have an oxyacetylene torch outfit, do not feel compelled to switch.

To torch-weld aluminum successfully, begin with an alloy like 3003, 6061, or 5052, which are all easily welded. Unweldable alloys such as 7075 and 2024 will drive you to drink because they inevitably crack after welding. For comparisons of the strengths and weaknesses of various alloys, visit the Aluminum Association website: www.aluminum.org. For an excellent improvised filler rod, cut a strip from the metal you intend to weld or use 4043 or 1100 filler rods. The latter are almost pure aluminum and will be softer than 4043. Both of these rods are widely available and

New aluminum will be identified when purchased, of course, but new aluminum is costly. Unfortunately, scrap aluminum is not always so clearly identified as this sheet of 2024, an unweldable alloy.

weld beautifully with a number of alloys, so you won't have to worry about incompatibility between your materials. You can find answers to your most arcane aluminum filler rod questions by consulting an aluminum filler alloy chart (see www.alcotec.com for one example), which lists different combinations of alloys and filler rods and compares them for different attributes: weldability, strength of the welded joint, ductility, corrosion resistance, ability to withstand sustained temperatures above 150 degrees Fahrenheit, and color match after anodizing. With the help of the chart, you can select rods having strengths in those areas that matter most for your projects. Strengths in some areas typically involve making a tradeoff somewhere else. I have a big spool of 1100 MIG wire for odd jobs; it is handy, and I know the welds will be workable.

Once you have the right alloys and rods, you need the right eyewear. When you try to torch-weld aluminum with conventional welding goggles, the incandescence of the flux makes you feel as if you are staring at a flare—the weld is completely obscured by a dancing orange flame. I know of two makers of welding goggles specifically designed to filter out the orange portion of the spectrum: TM Technologies (www.tinmantech.com) and Fournier Enterprises (www.

fournierenterprises.com). They are more expensive than conventional lenses, but the difference in enhanced visibility is dramatic. One immutable truth of welding is that you must be able to see what you are doing. According to the TM Technologies website, the question of the appropriate eyewear for aluminum welding is one of safety, and the site discusses the dangers of obsolete cobalt blue and didymium lenses that were once worn by glassblowers.

Torch selection and setup make a huge difference on how successful your aluminum welding is. For a point of reference, I use a Victor 00 for 0.036-inch steel and a Victor 0 for 0.063-inch aluminum. The latter welds as though it were made for this application—simply fantastic. I use roughly 3 to 5 pounds of pressure for the oxygen and the acetylene, but rather than turning up the pure acetylene flame until the soot burns off, I leave the acetylene flame soft. Moreover, you can have a lot of heat control simply by changing how intense your acetylene flame is. If you try to establish your first tack and it takes too long, turn off the oxygen and boost the acetylene a bit, for example. An acetylene flame that is 2 inches longer than before, but not so intense that it stops smoking, may make the difference between a flame that is too cool and a flame that is just right.

To begin welding, clamp some scrap aluminum sheet together just as you would with steel. Clean the weld area with denatured alcohol and a stainless-steel brush devoted to this purpose. Mix the powdered aluminum welding flux with water until it is about the consistency of milk. Gently heat the panels you plan to weld and run a flame over the end of your filler rod. Now paint the flux on the filler rod and on both sides of your panels. With just the right amount of heat, the flux will stick like the glaze on a day-old donut. To liquefy the flux, gently play a slightly carburizing torch flame over the work piece and the filler rod at a distance. If you heat them too abruptly, the flux will flake off and not do its job. Approach the seam as you would with steel, but probe the surface of your puddle with the end of your filler rod to see if it is molten. This practice will help prevent burning a hole. When the extreme edges of the metal on either side of the seam start to wrinkle ever so slightly, dab the spot with the filler rod, establish a tack and remove the heat. The little wrinkle that forms at the edges of the panels is the foolproof indicator that the panels are ready to be fused. From here forward, make a series of tacks about ¾ to 1 inch apart and then run a bead. Do not expect the parent metal to hold together quite as well while you weld as steel, so don't get carried away swirling your puddle or you will

Always prepare your flux according to the directions. In this case, you mix it with water to match the consistency of milk. Heat the metal gently and brush it over both sides of the seam. The panel will hiss, but the flux should stick. If the metal is too hot, the flux will jump off. It is a good idea to apply more flux after tacking the seam as well.

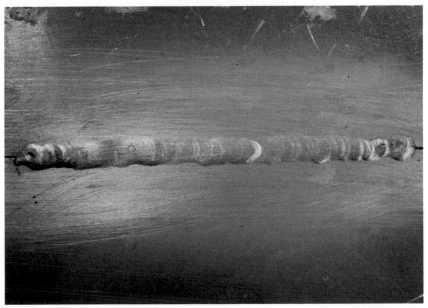

My aluminum torch weld on 0.063-inch-thick sheet is not as pretty as a TIG weld, but it is strong and much more ductile because of the high heat input of this process.

The back side of the aluminum torch weld in the previous illustration shows good penetration.

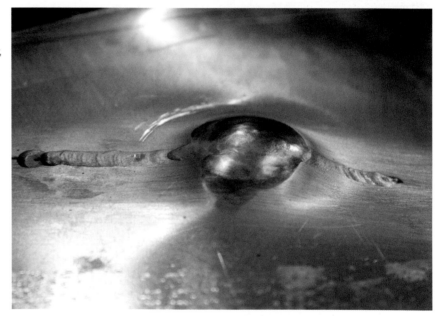

Aluminum torch welds are incredibly ductile—this one withstood repeated hammering in a planishing hammer without cracking. The ductility results from the annealing that takes place along the weld seam, however, rather than because of a mysterious quality inherent to torch welding.

burn a hole for sure. Just move the torch straight down the seam, dabbing the filler rod at the leading edge of the puddle as you go. You will notice that the puddle gets flatter as it gets hotter and, therefore, when it gets close to burning through. When that happens, lift the torch momentarily to allow the weld to cool slightly. Aluminum doesn't offer the same resistance to the filler rod—it's mushier—so it will take a little practice to get accustomed to the softer surface. Once the sides of the seam start fusing with the rod, move the puddle forward more rapidly than you would with steel. You may find it easier to control the heat if you reduce the angle between the torch and the work. When

you have finished your weld, thoroughly rinse the flux residue off with water to prevent corrosion and scrub it with a stainless-steel brush. If you find that you have a tendency to burn away the metal along the seam, readjust your flame with less acetylene or try clamping a strip of copper behind to support the metal. Lastly, do not expect an aluminum torch-weld to be as pretty as a TIG weld. With a lot of practice it can be, but torch-welding aluminum is a unique skill that develops somewhat independently. As long as your weld is solid, you can file it down to the point that no one will be able to tell that the weld wasn't as pretty as you had hoped.

## MIG WELDING (METAL INERT GAS WELDING)

In MIG welding, also known as gas metal arc welding, a thin consumable wire electrode is fed into the work piece through a hand-held gun to create a molten weld pool. The wire thus doubles as the filler rod. Although there are a number of wire-fed processes, we will concentrate on short-circuit transfer gas metal arc welding because it is the process most useful to you in sheet metal fabrication. Generally speaking, MIG is the easiest welding process to learn, it works well for welding

The relic at top left looks like something from a 1950s science fiction movie. Do not use gold lenses like this one if they have any scratches. The helmet at top right is inexpensive and functional. It's also dark all the time. The helmets at bottom are both auto-darkening, but differ vastly in price. In my experience, the inexpensive auto-darkening helmets are an excellent value. If ever they become sluggish, set them out in the sun to revive them.

This is a typical entry-level MIG welder, as used for welding sheet metal. A spool of wire that serves as the electrode and filler metal is fed into the work piece through the gun. A clamp attached to the work provides an electrical ground. Shielding gas, also fed through the torch, protects the weld bead from atmospheric contamination.

thick and thin metals, it is fast, and the entry-level machines are not terribly expensive. The only drawbacks to using the MIG for sheet metal are that the weld bead is hard compared with a torch or a TIG weld and you are limited to welding steel unless you buy additional equipment. MIG welders also tend to lay down a lot of metal on the bead, which can be a benefit if there is a gap to fill or an annoyance if you don't want to spend time grinding a large weld bead. The only new piece of personal protective equipment you will need for MIG welding is an auto-darkening welding helmet. You can certainly survive with the old style of helmet that is always dark, but the auto-darkening feature is indispensable when you need maximum accuracy.

As alluded to previously, MIG welding commences when the weldor squeezes the trigger on the MIG gun, simultaneously feeding the positively charged wire electrode into the negatively charged work piece and initiating the flow of shielding gas from the gun's tip. The advancing electrode arcs against the work and creates a small weld pool. Resistance heats up the end of the wire and the work piece until the surface tension of the weld pool and an induction-related pinch force pull the molten metal from the end of wire. The wire is, of course, fed back into the work, arcs, and melts, and the process repeats several times a second. When welding, these short circuits happen so rapidly that the sound created is simply a steady sizzle or crackling. The molten weld pool is protected from atmospheric contamination by the shielding gas, usually carbon dioxide or a mixture of argon and carbon dioxide.

For satisfactory results, select your MIG wire based on the type and thickness of your base metal. Larger wire carries more amps and thus more heat, so your wire diameter will most likely not mirror the thickness of your metal in the manner that torch filler rods do. I recommend the smaller wire for thin sheet metal. We use 0.025-inch diameter ER70S-6 wire in the small MIG welders in our school shop. Other common sizes are 0.030 inch and 0.035 inch. The ER70S-6 is an American Welding Society (AWS) specification for carbon steel filler metal used in gas metal arc welding. The ER stands for *electrode/rod*, meaning that this filler metal is the same specification whether it is contained in a wire electrode or a filler rod used in gas or TIG welding. The S is for *solid* wire,

as opposed to tubular flux-core. The 70 represents 70,000 psi tensile strength, and the 6 identifies this wire as one of seven clearly defined versions of this filler rod based on chemical composition. ER70S-6 is a very common low-carbon wire that works well in automotive restoration applications. You can get small 5-pound spools of different wires if you wish to experiment.

The most popular shielding gases for MIG welding steel are carbon dioxide and a mixture of the former with argon. In heavy plate steel, carbon dioxide allows for deeper penetration, but with thin sheet steel you will prefer the cleaner weld of the argon/carbon dioxide blend. Straight argon is usually used with aluminum, whereas a mix of carbon dioxide, argon, and helium is used for stainless steel. The carbon dioxide/argon blend also allows you to weld 4130 steel. If you choose pure carbon dioxide as your shielding gas, get a regulator designed for it. Not all regulators may be used with more than one gas. Consult your gas supplier if you have special requirements.

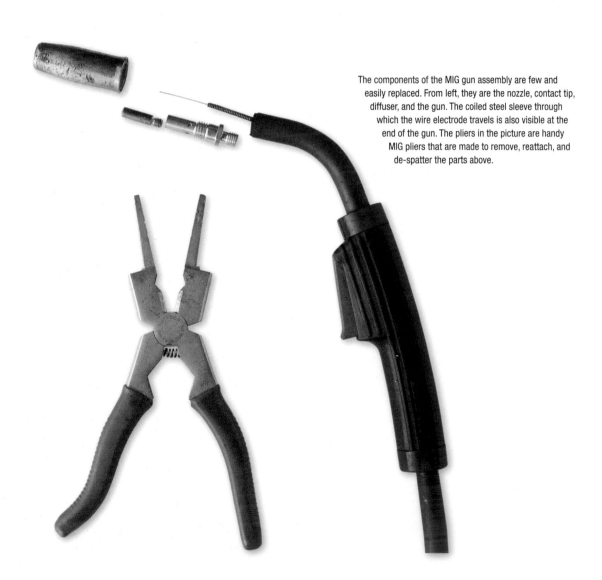

The components of the MIG gun assembly are few and easily replaced. From left, they are the nozzle, contact tip, diffuser, and the gun. The coiled steel sleeve through which the wire electrode travels is also visible at the end of the gun. The pliers in the picture are handy MIG pliers that are made to remove, reattach, and de-spatter the parts above.

This MIG regulator indicates your cylinder pressure and controls the flow of shielding to the gun. Note that it has two scales depending on the gas used.

Many entry-level wire feed welders are sold as flux-core wire arc welders that can be upgraded to MIG operation by adding a shielding gas cylinder, swapping out the flux-core wire for solid wire, and attaching the regulator that is included with the machine. Flux-core wire is a hollow wire with flux contained inside to create an inert atmosphere around the weld pool during welding, rather than through a separate shielding gas. The flux-core wire welders produce a lot of smoke and weld spatter, but they work. I have one reservation regarding flux-core wire, however. Flux-core wire will not work well for the MIG welding technique for sheet metal that I will describe shortly because flux will become imbedded in the weld. As a cautionary note, these welders usually come set up for bigger wire than you will want for solid MIG wire welding of thin sheet metal. Fortunately, the wire feed drive wheels can often be reoriented to fit 0.025-inch or 0.030-inch solid MIG wire or 0.035-inch flux-core wire. Before purchasing one of these welders, you might investigate what wire sizes the equipment will handle and what might be required to change from one size to another. Some things to consider would be matching contact tips for the new size wire, possibly a new gun liner, and you may have to change polarity to DCEP from the DCEN that flux-core wire often requires, a simple swap of two wires above the MIG wire feed mechanism on the side of the machine.

If you are shopping for an entry-level welder, the smaller 110-volt welders will work well for thin sheet metal. These are particularly versatile because they will work with your house current. If you plan to weld steel thicker than about 14 gauge, you might select a 200/230-volt welder, which will mean less flexibility regarding where the welder can be used, but it will still perform nicely on thinner metals.

A second type of MIG regulator, or flowmeter, also with two scales.

This is the view under the side cover of a MIG welder. The wire spool easily pops on and off. The wire electrode is fed between a drive wheel and an unpowered idler wheel on this machine. The spool is labeled to identify the type of filler metal.

To set up the machine, load a spool of wire in the welder, feed the loose end between the wire drive wheels, turn the power on, and squeeze the trigger on the gun. Now turn the wire speed button on the front of the machine all the way up and wait for the wire to exit the gun nozzle. When the wire appears, turn back the wire speed knob, point the gun at an ungrounded table top or kneel down and point it at the floor. With the nozzle an inch or so from a hard surface, continue pulling the trigger to see if the wire will overcome the resistance you've just introduced. If the wire stops advancing, tighten the wingnut that applies tension to the wire feed drive wheels a turn or two to prevent the wire from slipping. Do not overtighten it to the point that the wire is mashed between the wheels. Attach your shielding gas regulator to its cylinder and follow the same procedure you would with an oxygen tank. Set your low-pressure gauge to between 20 and 30 CFH (cubic feet per hour). Check the chart on the inside of the machine for the approximate voltage and wire speed settings for the

gauge of sheet metal you want to weld and adjust the welder accordingly. Attach the work clamp to your table or work piece and run a few practice beads on a piece of metal at least 1/8-inch thick. Running a bead is as easy as placing the nozzle of the gun about 1/8 inch from the work piece, pulling the gun trigger, and moving the puddle along steadily in a forehand manner so that you deposit a nice, even bead of weld. You will probably want to lean over to the side so you have a clear view of the weld zone as you go. Do not situate yourself so that the gun blocks your view of the weld. I find that the recommended settings are typically not hot enough for my liking. Do you hear a resounding sizzle when you weld with the recommended settings or does it sound a little anemic? Boost the voltage and wire speed up to the next highest setting and try again. For the welding technique I recommend, you want plenty of heat for maximum penetration. Turning up the voltage will tend to flatten and widen the bead, whereas increasing the wire speed will deposit more metal in the weld.

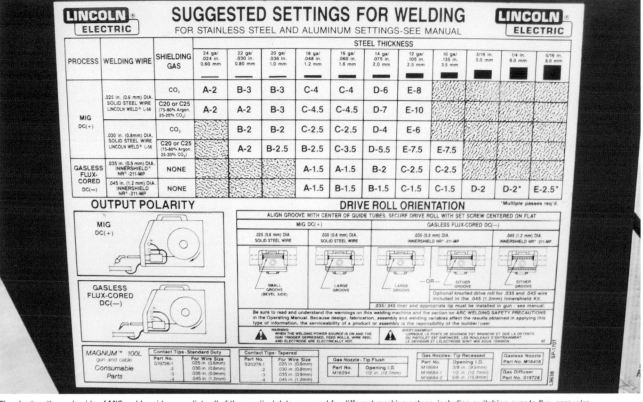

## SUGGESTED SETTINGS FOR WELDING
### FOR STAINLESS STEEL AND ALUMINUM SETTINGS-SEE MANUAL

| PROCESS | WELDING WIRE | SHIELDING GAS | 24 ga/ .024 in. 0.60 mm | 22 ga/ .030 in. 0.80 mm | 20 ga/ .036 in. 1.0 mm | 18 ga/ .048 in. 1.2 mm | 16 ga/ .060 in. 1.6 mm | 14 ga/ .075 in. 2.0 mm | 12 ga/ 105 in. 2.5 mm | 10 ga/ .135 in. 3.5 mm | 3/16 in. 5.0 mm | 1/4 in. 6.0 mm | 5/16 in. 8.0 mm |
|---|---|---|---|---|---|---|---|---|---|---|---|---|---|
| MIG DC(+) | .025 in. (0.6 mm) DIA. SOLID STEEL WIRE LINCOLN WELD® L-56 | CO₂ | A-2 | B-3 | B-3 | C-4 | C-4 | D-6 | E-8 | | | | |
| | | C20 or C25 (75-80% Argon, 25-20% CO₂) | A-2 | A-2 | B-3 | C-4.5 | C-4.5 | D-7 | E-10 | | | | |
| | .030 in. (0.8mm) DIA. SOLID STEEL WIRE LINCOLN WELD® L-56 | CO₂ | | B-2 | B-2 | C-2.5 | C-2.5 | D-4 | E-6 | | | | |
| | | C20 or C25 (75-80% Argon, 25-20% CO₂) | A-2 | B-2.5 | B-2.5 | C-3.5 | D-5.5 | E-7.5 | E-7.5 | | | | |
| GASLESS FLUX-CORED DC(—) | .035 in. (0.9 mm) DIA. INNERSHIELD® NR® -211-MP | NONE | | | | A-1.5 | A-1.5 | B-2 | C-2.5 | C-2.5 | | | |
| | .045 in. (1.2mm) DIA. INNERSHIELD NR® -211-MP | NONE | | | | A-1.5 | B-1.5 | B-1.5 | C-1.5 | C-1.5 | D-2 | D-2* | E-2.5* |

*Multiple passes req'd.

### OUTPUT POLARITY

MIG DC(+)

GASLESS FLUX-CORED DC(—)

### DRIVE ROLL ORIENTATION

ALIGN GROOVE WITH CENTER OF GUIDE TUBES, SECURE DRIVE ROLL WITH SET SCREW CENTERED ON FLAT

MIG DC(+) | GASLESS FLUX-CORED DC(—)

.025 (0.6 mm) DIA. SOLID STEEL WIRE — SMALL GROOVE (BEVEL SIDE)

.030 (0.8 mm) DIA. SOLID STEEL WIRE — LARGE GROOVE

.035 (0.9 mm) DIA. INNERSHIELD NR® -211-MP — LARGE GROOVE — OR — EITHER GROOVE

.045 (1.2 mm) DIA. INNERSHIELD NR® -211-MP — EITHER GROOVE

Optional knurled drive roll for .035 and .045 wire included in the .045 (1.2mm) Innershield Kit.

.035/.045 liner and appropriate tip must be installed in gun - see manual.

Be sure to read and understand the warnings on this welding machine and the section on ARC WELDING SAFETY PRECAUTIONS in the Operating Manual. Because design, fabrication, assembly and welding variables affect the results obtained in applying this type of information, the serviceability of a product or assembly is the reponsibility of the builder/user.

**WARNING:** WHEN THE WELDING POWER SOURCE IS ON AND THE GUN TRIGGER DEPRESSED, FEED ROLLS, WIRE REEL AND ELECTRODE ARE ELECTRICALLY HOT.

**AVERTISSEMENT:** LORSQUE LE POSTE DE SOUDAGE EST BRANCHE ET QUE LA GACHETTE DU PISTOLET EST ENFONCEE, LES ROULEAUX D'ENTRAINMENT, LE DEVIDOIR ET L'ELECTRODE SONT MIS SOUS TENSION.

MAGNUM™ 100L gun and cable — Consumable Parts

| Contact Tips - Standard Duty | |
|---|---|
| Part No. | For Wire Size |
| S19726-1 | .025 in. (0.6mm) |
| -2 | .030 in. (0.8mm) |
| -3 | .035 in. (0.9mm) |
| -4 | .045 in. (1.2mm) |

| Contact Tips - Tapered | |
|---|---|
| Part No. | For Wire Size |
| S20278-1 | .025 in. (0.6mm) |
| -2 | .030 in. (0.8mm) |
| -3 | .035 in. (0.9mm) |
| -4 | .045 in. (1.2mm) |

| Gas Nozzle - Tip Flush | |
|---|---|
| Part No. | Opening I.D. |
| M16294 | 1/2 in. (12.7mm) |

| Gas Nozzles - Tip Recessed | |
|---|---|
| Part No. | Opening I.D. |
| M16684 | 3/8 in. (9.5mm) |
| M16684-1 | 1/2 in. (12.7mm) |
| M16684-2 | 5/8 in. (15.9mm) |

| Gasless Nozzle | |
|---|---|
| Part No. M16418 | |
| Gas Diffuser | |
| Part No. S19720 | |

The chart on the underside of MIG welder side cover lists all of the practical data you need for different machine setups, including switching over to flux-core wire.

You can't weld if you can't see what you are doing. Lean over to keep the weld pool in view.

Once you feel that the machine is properly set up to run a continuous bead on thicker metal, clamp two pieces of your chosen sheet metal together in anticipation of creating a butt joint. Leave a gap between the pieces equal to a little more than the thickness of the metal. Too little gap and the pieces will butt together and create a hump along the weld seam; too much gap will require an excessive amount of filling. Establish a row of tack welds along the seam about an inch apart. The gun should make a satisfying *brap* sound with each pull of the trigger as the individual tack welds instantaneously penetrate the seam. If the machine is not hot enough, the weld will drizzle along the surface, failing to penetrate completely. Alternate the placement of the tack welds so that you are not heating up the metal too much as you go. Keep the metal on both sides of the seam in the same plane at all costs. Don't be afraid to cut tack welds with a cut-off wheel if necessary to realign the panels you are welding if they wander out of the same plane. Use your free hand to steady the gun or to act as a rest for your gun hand. When the tack welds are complete, start on one tack and perform three or four tacks in succession, each slightly overlapping the one that came before it. Move to a different area on the seam and repeat this operation until the entire seam is welded. If you have stayed in control of the weld the entire time, the panels will still be in alignment. As an additional precaution to prevent heating up the panel, try the method just described, but blast each series of welds with compressed air from a blow gun as soon as the color leaves the last weld in each group. Let the air cool the welds for about four to five seconds each time to minimize heat distortion—an old body shop technique that works well. You may have some slight shrinkage in the panel to deal with, but I will discuss that in connection with metal straightening in a later chapter.

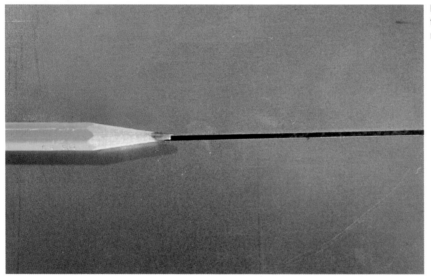

My rule of thumb for the MIG is to leave a gap between the panels equal to approximately 1½ times the metal thickness.

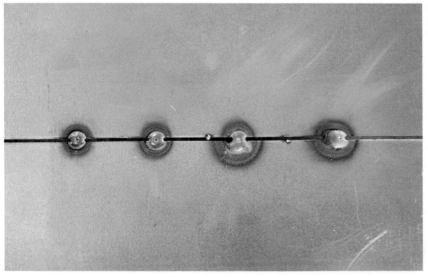

Establish your tacks about ¾ to 1 inch apart and maintain panel alignment at all costs. Setup is probably 90 percent of a successful weld.

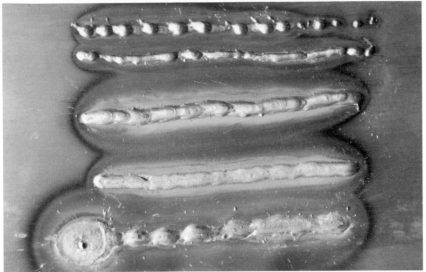

Use these examples for reference when dialing in your MIG welder. From top to bottom: 1) wire speed and voltage are too low—the weld is poorly fused to the parent metal; 2) the voltage has been turned up, but still no penetration; 3) the wire speed has been turned up to keep up with the increased voltage in No. 2, but more penetration is needed; 4) this looks good; 5) increasing voltage farther is too hot, as evidenced by extra-wide, flat bead.

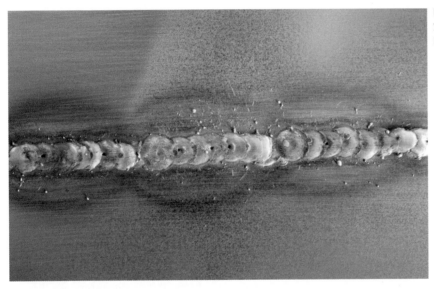

For thin steel, set up your welder pretty hot and overlap your welds in short bursts of three, moving around the panel so as not to overheat one area.

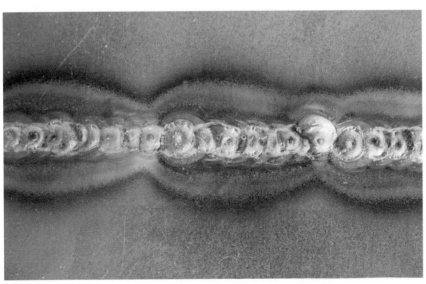

Back side of weld in previous illustration shows complete penetration. Although this weld would require some grinding on the top side, it could be sanded to a very smooth finish.

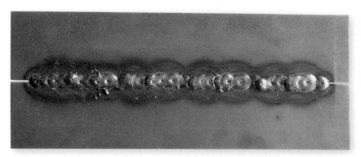

To reduce heat warpage even further, try the overlapping weld method described in the text and hit each series of welds with compressed air as soon as the color leaves the last weld in each group.

Because MIG wire serves as both the electrode and the filler rod, MIG welding works great for filling unwanted holes. Some holes can be filled with a few brief bursts from the gun, allowing a moment's hesitation between each new weld to allow the molten metal to solidify in the hole. For larger holes, clamp a piece of copper behind the hole or hammer the end of a length of copper pipe flat and use it for backing up inaccessible holes. Two or three quick bursts from the gun and the hole will be filled.

After several years of using MIG welders heavily in auto restoration, I'm convinced they are some of the most durable machines made. The most common problems I have seen are gun nozzles that accumulate weld spatter and interfere with proper gas shielding, ground cables that become broken from having the welder rolled over them, and occasionally

a gun liner that does not feed properly due to an internal kink from having been bent, run over, or coiled too tightly. Nozzle spatter is easily avoided by keeping the gun clean or by using one of the various nozzle dip products that can be applied to the end of the gun to hinder spatter from sticking. Broken ground cables are deceiving because the copper wire inside the cable may be broken even though the insulation is intact. If the machine seems to run fine but won't arc, inspect the ground. Check the grounding cable for continuity with an ohmmeter. If the wire hangs up during feeding, check the wire drive rolls, the contact tip, and the diffuser. If you've checked the wire spool, checked the drive rolls, and removed the contact tip and the diffuser and the feed is still intermittent, the wire is probably hanging up inside the gun liner, which is easily replaced.

When it comes to filling holes, the MIG is king. Do you really want to make a patch for this ⅜-inch hole and then try to weld it in without the patch moving?

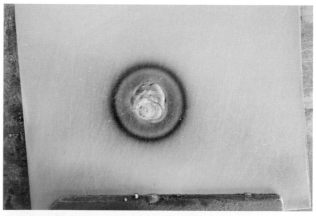

This hole was too small to make a patch for, so I clamped a piece of copper behind it and used the MIG welder to fill the hole with two or three quick bursts of weld.

## TIG WELDING

If you like torch-welding, you will probably love TIG welding. The torch manipulation is very similar between the two, and you will find that the manual skills you develop with oxyacetylene welding translate directly to the TIG. Most TIG enthusiasts appreciate the superior heat control that TIG welding offers by way of a foot- or thumb-operated remote amperage control. TIG welding machines are more costly than starter oxyacetylene torch setups or starter MIG welders, but after a few minutes of using one you will realize that you have entered a new realm of precision weld control.

TIG gets its name from its nonconsumable tungsten electrode and the inert shielding gas that is used in this process. Formerly, helium was the prevailing shielding gas, rather than the argon of today, so don't be confused by the older term heli-arc welding. Helium is still sometimes used as a shielding gas, but not nearly as often as argon. TIG is usually referred to as GTAW (gas tungsten arc welding) in industry literature. The main benefits of TIG welding for our purposes are that it does not put a lot of heat into the work piece, it will weld a variety of metals, and the finished weld is soft and easy to shape. Besides the high cost of the machine, the only drawback to TIG welding is that it is very slow. I remember when I worked at a restoration shop restoring British cars, I typically only used the TIG for aluminum or when I was welding together pieces that needed additional shaping. It was painfully slow compared with the MIG. Speed might matter if you have production requirements or you are billing out your time.

TIG welding works by creating an electric arc between the nonconsumable tungsten electrode in the torch and the grounded work piece. You start the arc and control the amperage with a remote foot pedal or thumb control on most modern machines. The heat from the arc creates a weld

The Miller Synchrowave 180 SD at left was once the common size among entry-level welders, but in recent years the size and cost of TIG machines has shrunk. The Miller Maxstar 150 STH is DC only, but it's the size of a lunchbox and as cute as a button.

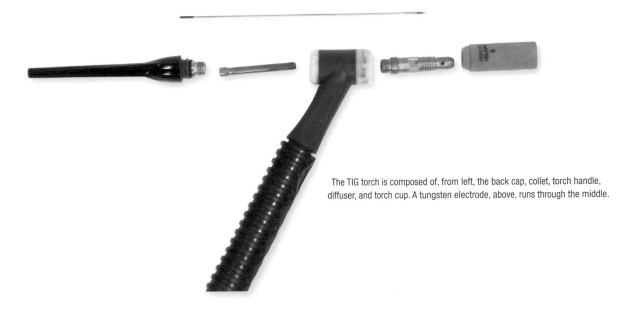

The TIG torch is composed of, from left, the back cap, collet, torch handle, diffuser, and torch cup. A tungsten electrode, above, runs through the middle.

puddle that you can move along the seam and add filler rod with your other hand as needed. The tungsten electrode protrudes from a ceramic cup that directs an inert shielding gas over the weld area. The cups are numbered according to size, the bigger ones providing better gas coverage at the expense of maneuverability in confined spaces. The rear end of the tungsten is covered by a back cap that prevents the tungsten from accidental arcing. As with torch cups, back caps are available in different sizes for the sake of maneuverability.

In the interest of brevity, I cannot provide an exhaustive list of features that are available for TIG machines. Seemingly every aspect of the machine's operation may be fine-tuned in detail. Naturally, cost increases with enhanced control and expanded features. I will focus, therefore, on those elements necessary for sheet metal welding.

TIG welders are expensive, but they are much less so than in the past, thanks to advances in technology. The machines are also more compact than ever. Before acquiring a TIG welder, think carefully about the type of metal you want to weld. If you work with aluminum, you want a TIG welder with AC capability. Remember those annoying oxides that form on the surface of aluminum? Alternating current breaks up those oxides and greatly facilitates aluminum welding. For most other welding, you will use DC electrode negative. An air-cooled torch will suffice for most sheet metal work where you will be dividing your time between shaping and welding. If you spend a lot of time welding, you may need a water-cooled torch to avoid having to stop to allow the torch to cool.

## TUNGSTEN ELECTRODES

Although you may choose from many types of electrodes, two types will take you a long way in TIG welding: the pure tungsten and the 2 percent thoriated tungsten electrodes. Each is color-coded: usually with a green stripe for pure tungsten and a red stripe for thoriated, depending on the manufacturer. You may wish to choose an alternative to thoriated tungsten, however, because it is a radioactive material and may, therefore, be hazardous to your health. Though it poses little danger as a whole electrode, its dust is more of a threat because it can be more easily taken into the body in this form. Discuss your concerns with your welding supplier, and conduct research to determine the threat it may pose to you. If you proceed with thoriated tungstens, use them for ferrous metals and use pure tungsten for aluminum. Each type of tungsten has to be prepared for welding in a specific way. Thoriated tungstens must be ground lengthwise to a sharp point to obtain a focused arc, while pure tungstens ball up on the end after a few moments of use to create the wide arc needed for welding aluminum. If you touch the electrode with the filler rod or if you touch the work piece with the electrode, remove and regrind the tip of the tungsten to prevent contaminating the weld. Filler rods melt and cling to the tungsten with sickening efficiency, especially when the electrode is freshly sharpened. If the accumulation of filler on the tungsten is large, crack off the hideous lump between two pairs of pliers and resharpen the tungsten. Point the tip of the tungsten down and rotate it between your fingertips against a bench grinder or belt sander devoted

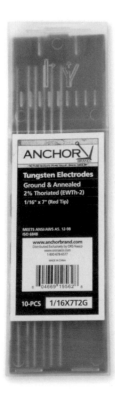

Tungstens are color-coded for easy identification once they've left the package, but not every maker uses the same color scheme. In this case, the 2 percent thoriated tungsten is red, and the pure tungsten is green.

to this purpose. For most welding, adjust the tungsten so it protrudes about ⅛ inch from the torch cup. If the tungsten doesn't stick out enough, you won't be able to see what you're doing. If it sticks out too much, the tungsten will not be adequately shielded and will get too hot.

To begin welding, practice on steel using a thoriated electrode ground to a point. Set the machine to DCEN, and attach the ground clamp to your worktable or the piece itself. Set up your gas cylinder and regulator just as you did with the MIG, and adjust the low-pressure gauge to around 20 CFH. Depending on the machine, you may hear a hissing sound as soon as you turn it on; this is the shielding gas preflow which is intended to protect the weld zone in anticipation of welding. The preflow and postflow, as it is called after the weld is complete, are adjustable features on the more sophisticated machines. Holding the torch at about a 60 degree angle to the work, bring the tip of electrode to within about ⅛ inch of the panel and gently step on the

When ready for use, the pure tungsten forms a ball on the end while the thoriated tungsten must be sharpened to a point.

Hold your thoriated tungsten down against a grinder or belt sander that you use for this purpose only. Rotate it in your fingers to create a nice point for a focused arc. Beware of grinding dust created in this process. Installing a dust collection system would be a good idea.

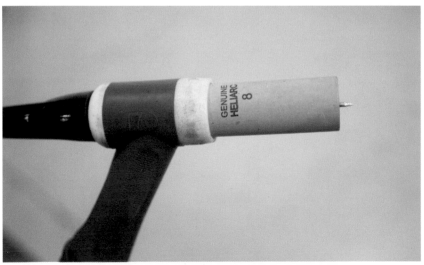

For most sheet metal TIG welding, secure the tungsten electrode so that it sticks out ⅛ inch past the torch cup.

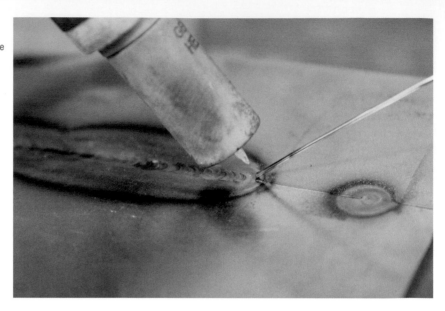

The process of TIG welding, ironically, is a lot like torch welding. Create a puddle, move it along at an even rate, and dab the rod into the leading edge of the puddle. At the completion of each TIG weld, hold the torch in place for several seconds to allow the shielding gas to protect the solidifying weld pool.

remote amperage pedal or turn the thumb dial on your torch. Experiment with different torch angles to find what works for you. Different angles create arcs with different characteristics. At first the arc will be small and wavering and will make a crackling sound. As you increase the amperage, however, it will stabilize and become quiet, and a small molten pool will form. You will notice that increasing the amperage widens the pool until you burn a hole. Practice creating a pool and moving it across the work surface.

If you have torch-welded, adding the filler rod will be second nature. As you probably realize from the earlier discussion of filler rods, there is a rod for every purpose. A good choice for steel is the commonly available ER70S-2, or you can recycle clippings from your MIG welder if you have one. You can use these for gas welding, too, but avoid using the gas rods such as RG 45 with the TIG other than for practice because they will not weld as cleanly. Simply move the pool along steadily and dab the rod at the leading edge of the molten pool. There is no need to move the electrode forward and back as you may have seen in diagrams of TIG welding. Just move the torch forward at a steady rate and at a steady distance from the piece. Keeping the filler rod close to the arc will keep it hot and within the protective envelope of the shielding gas. You will no doubt touch the rod to the electrode from time to time. Simply stop, regrind the electrode, and continue welding. After some practice, you will very seldom have this problem. You may also stick the electrode to the piece. This, too, will require removing and regrinding the electrode. At the end of each weld, reduce the heat slowly by gradually backing off your amperage control, and leave the torch over the weld for a few seconds to allow the postflow of shielding gas to do its work; doing so allows the weld to cool slowly without becoming contaminated. TIG welding is such a tranquil activity that it is easy to forget that you are generating ultraviolet rays that will

damage your skin. Cover unprotected areas because you will not notice burns until well after you are finished.

Butt joints with the TIG do not require the gap that MIG welds do. Often, you can get away with lightly butting the edges of the work pieces together. Have the filler rod poised near the torch cup in the event the edges of the metal along the seam start to burn away as you add heat, rather than fusing together. Perform a series of tack welds, hammer out or otherwise correct any misalignment issues, and begin connecting your tack welds. Try to place your welds as far from one another as possible as you go to prevent heating up the panels excessively.

## TIG WELDING ALUMINUM

For aluminum, set the machine to AC to help break up the formation of surface oxides, and swap out the thoriated tungsten in your torch for one of pure tungsten. Thoroughly clean the surface to be welded just as you did with the torch. You will notice that the AC arc sounds even angrier at first than does the DC, but it, too, will settle down as you increase the amperage. I find that holding the torch almost perpendicular to the work makes a nice hot death ray that I can direct with more control than when the torch is canted. If the tungsten is a new blunt one, simply creating an arc and moving it along over the metal will create the proper ball on the end of the electrode. Without a filler rod, the weld puddle on aluminum will be noticeably concave, like a trough. Therefore, when TIG welding aluminum, use a filler rod. Starting with either a 4043 or 1100 rod as you did with the torch, steadily move the puddle along and dab the rod at the leading edge of the puddle. Do not weave the torch in any way; just move it along in a forehand manner. If the tip of the rod wanders out of the shielding gas envelope while it is hot, an oxide-covered ball will form on the end of the rod.

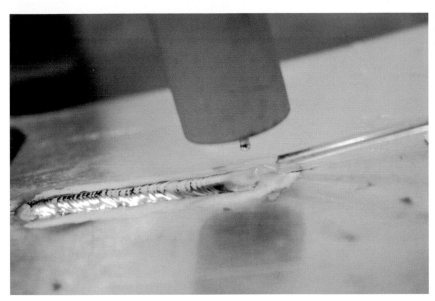

Experiment with different torch angles between about 60 and 90 degrees. Close to perpendicular works well on aluminum.

If you notice this happening, flick it off while the ball is hot or simply stop and cut off the contaminated end of the rod. Finding the right distance between the rod and the puddle can be tricky, but it helps to plan to use the rod the entire time and keep it close to the torch cup. Hopefully, with the filler rod close to the action and in continuous use, you won't have a problem with it balling up.

Executing a TIG-welded butt joint on aluminum is only slightly trickier than on steel. Butt two panels together and clamp them. As soon as you strike an arc on the joint, the edges of the panels will behave like feuding cousins who cannot stand to be in one another's presence. The more heat you apply, the more quickly the edges retreat from one another, leaving a large hole behind. The key is to bring in plenty of heat right away so the edges become shiny—this is the equivalent of puddling, but the edges are still separate—and introduce the rod immediately. The filler rod disrupts the surface tension of the molten pools on the panel edges and allows them to flow together. Once a single puddle forms between the two panels, you can increase the amperage as needed to fuse the metal at that spot. Be particularly careful about slowly decreasing your amperage as you finish this and every aluminum TIG weld. If you reduce your amperage abruptly, the weld will always crack. Once you have established a series of tacks, connect your tack welds. To keep the ends of the panel from burning away as you approach the end of a weld seam, clamp a piece of copper behind and plan to decrease the amperage as you near the edge.

The balance control allows you to boost the current as it leaves the work piece. This feature can be helpful when trying to weld aluminum with stubborn surface oxides.

If you find yourself working on aluminum that is unreceptive to welding, it is either one of the nonweldable alloys, or, more likely, the surface is dirty or corroded. Any foreign material on the surface, whether an oxide or otherwise, will become a dark weld-proof scab under welding heat. If the metal is slightly corroded despite your best efforts to clean it, experiment with the balance control if your welder is equipped with this feature. The balance control boosts the current as it comes off the work piece as an aid to breaking up surface oxides. You will increase the cleaning action at the expense of penetration. I found the balance feature helpful when working on dilapidated British cars.

## TIG WELDING STAINLESS STEEL

TIG welding stainless steel is essentially the same as welding mild steel except that you need a stainless filler rod and you need to protect the back side of the weld from atmospheric contamination that would otherwise weaken the weld. This protection may take the form of back purging with a shielding gas, or you can apply a specific type of flux to the back side of the weld seam. For back purging, you will need another argon cylinder and regulator. Create a small, shallow chamber out of poster board and tape to enclose the back side of the seam. This setup need not be beautiful or elaborate, just airtight. Run a hose from the regulator to the chamber, seal the entry hole with tape, and run a second hose, also sealed, out of the other end of the chamber. Set the low-pressure gauge to 20 CFH on your regulator and charge the chamber with argon. Run the exhaust outside so that you don't displace all the oxygen in your work area. With a shallow chamber on the back, you'll be ready to weld within moments. Setting fire to or otherwise damaging your chamber hinders its functionality, of course, so keep your tape handy.

To TIG weld stainless without a second argon cylinder and regulator, try a product called Solar Flux, which is used a lot in our area for industrial applications. You mix it with methyl alcohol to the consistency of cream and then paint it on the back side of the seam. After welding you can clean off the glassy residue on the back with a wire brush if you desire.

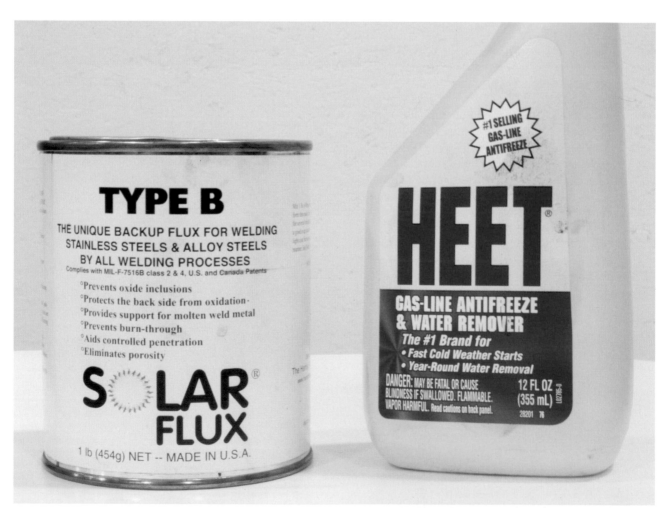

Mix the product Solar Flux with methyl alcohol, commonly available in Heet fuel line water remover, to create a flux for protecting the back side of your stainless TIG welds.

After the Solar Flux and methyl alcohol have been mixed to the consistency of cream, paint them on the back side of your seam.

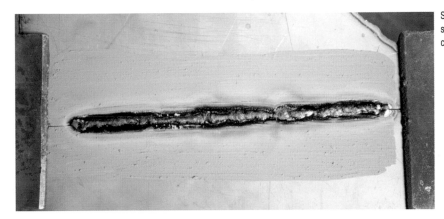

Solar Flux does not burn off during welding and can be scrubbed off with a stainless-steel brush after the weld is complete. Here is the back side of the weld after welding.

On thin metal, there is no need to slather the flux in the joint. In fact, if any of the flux oozes through to the top side from the application on the back, I recommend gently brushing it away before welding with your stainless brush. This photograph shows the top side of the finished stainless weld backed with Solar Flux.

This piece of stainless automotive trim had a crack and a chunk missing that I filled by coating the back with Solar Flux. A piece of thin brass adds further support to allow me to fill the void with stainless filler rod.

# Chapter 3
# Brazing, Soldering, and Riveting

In this chapter I will discuss three nonwelding fastening processes that are commonly used in the restoration of antique automobiles. Brazing and soldering rely on heat, of course, but they are different from welding in that the base metal is not melted and fused together. Instead, a filler metal forms a bond between two metals through surface adhesion. Riveting, on the other hand, is typically done cold, but you need to bring in a torch to install steel rivets successfully.

## BRAZING AND SOLDERING

Brazing is useful for repairing cast iron, for joining dissimilar metals, and for building up worn gear teeth and other surfaces. One of the purported benefits for brazing is less heat distortion due to the lower temperatures involved compared with torch-welding. Although there is no disputing the temperature difference involved between torch-welding and brazing, I am dubious of the distortion claim. The metal still must be brought to a red heat to braze and the metal is heated more slowly than in torch-welding. In addition, brazing requires an overlapped joint, so there are two thicknesses of metal to heat. As a result, I believe that more total heat ends up in the panel than with welding. Also, an overlapped joint is thick and hard to finish compared with a welded butt joint. I would never choose brazing over welding when dealing with thin sheet metal, but brazing is still useful for the other purposes previously cited.

Brazing rods are made of brass, an alloy of zinc and copper, and must be used with a flux, which facilitates the flow of the molten filler rod and promotes good adhesion by dissolving oxides and preventing atmospheric contamination.

The rivets on this Ford Model T frame must have a certain appearance if they belong to a car that is being restored to like-new condition.

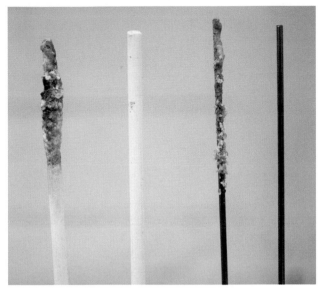

Here is a sampling of brazing rods and their associated fluxes. From left: flux-coated rod after use, a new flux-coated rod, a rod that has been heated and dipped in flux, and a bare brazing rod.

Fluxes are available in a powdered form that you dip the hot filler rod into, or you can buy precoated filler rod. The precoated rod is convenient, but the flux cannot be mistreated or it will break off, and the rods need to be carefully stored to prevent the coating from absorbing moisture. Always match your fluxes to your processes; fluxes are not interchangeable.

During brazing, the base metal is heated red hot (for steel), and the filler metal is touched to the surface. If the rod and base metal do not get hot enough, the braze tends to form a clump on the surface. On the other hand, if the

brazing material gets too hot, it swims across the surface, boils angrily, and gives off white smoke as the zinc is boiled away. The fumes given off by brazing fluxes and overheated brazing rod are bad for you, so always use adequate ventilation and keep your head out of the smoke. Because brazing relies on surface adhesion, you need joints with plenty of surface area, such as lap and strap joints, rather than butt joints. If you are building up material on a surface, as we will do in the demonstration piece, you won't need an overlap, however.

For brazing I usually use the same tip I use for sheet metal welding—a Victor 00, 0, or 1—unless the piece is really large, adjusted to a neutral flame. Because you are not trying to puddle the base metal, keep the torch tip back with the cone at least an inch or an inch and a half from the piece; you will be heating with the flame's envelope. With the metal very clean, play the torch over the surface to heat it up. As soon as the surface gets dull red, play the torch over the end of the rod, dip it in the flux, and apply it to the work piece. For a brazed joint, you will want to make a series of tacks just as if you were welding and then travel down the seam. For building up an area, heat it, apply the rod, allow the rod to solidify by cooling slightly, and repeat the process. Don't forget to keep applying flux to the rod as the rod is consumed.

Part of the steering linkage on our 1908 Holsman project had a worn oblong hole that needed to be restored. Brazing seemed like an expedient technique for making the repair. We heated the piece red hot, heated the tip of the brazing rod, and dipped it into the flux. Because the hole was oblong by about ¼ inch on one side, several applications of brazing rod were needed to build up the inside of the opening. When we felt that we had added enough material, we allowed the piece to cool and scrubbed off the flux residue with a wire brush and water.

The first step in brazing the Holsman steering component is to heat the part until it's red hot.

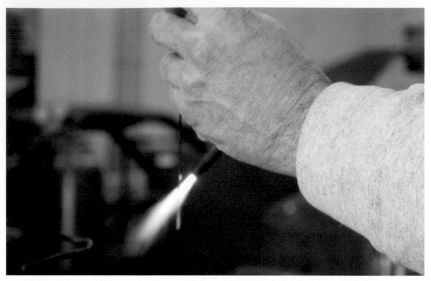

Heat the brazing rod and dip it into a can of powdered flux. The flux will stick to the rod.

Apply the brazing rod to the hot part. Allow some braze to pile up on the part, and then move the torch for a moment to allow the braze to solidify. Repeat as needed to build up the surface.

If the brazed part gets too hot, the zinc will boil out of the braze and you will see a cloud of white smoke and a lot of white residue, as here.

Here is the finished brazed repair after it has cooled and been cleaned off.

Like conventional brazing, silver soldering takes place at temperatures above 840 degrees Fahrenheit. We use silver solder occasionally in the restoration of antique cars to repair brightwork on brass-era cars. Silver solder is an alloy of silver, zinc, and copper and requires a flux for successful application. Silver solder can join copper, brass, stainless steel, and mild steel. For the demonstration, we silver soldered a piece of brass to a piece of steel. We started with clean metal. We applied liquid silver soldering flux and added small snippets of silver solder. The solder can be applied directly from the wire, but it is easier to avoid applying too much solder by using the small pieces. The flux helps hold the flux pieces in position while the work is being heated. Gradual heating helps keep the flux from jumping off and disturbing the solder. With brazing and soldering generally, maintain a little clearance between the pieces to allow the solder to climb between them by capillary action. If the work pieces are clamped too tightly, the filler metal cannot get into the joint. Heat the work, not the solder, just enough to make the silver solder become liquid and flow, and then remove the torch.

Unlike silver solder, traditional soft solders melt below 840 degrees Fahrenheit. Two soft solders we use frequently are acid-core tin/lead solders and tin/lead automotive body solder. Acid core solder contains an acidic flux to clean the base metal to which it is applied to promote solder adhesion. We use acid core in reattaching old soldered hard lines and for repairing containers. We use lead auto body solder to seal factory body seams and for repairs that will be subjected to vibration or high stress. I will discuss body solder momentarily.

To demonstrate silver soldering, we rounded up a piece of brass, some mild steel, some silver soldering flux, and wire silver solder.

We cleaned our metal, applied flux to the areas to be soldered, and lightly clamped our pieces together. Don't clamp the pieces too tight or the solder won't be able fill the joint. We cut our solder into several small pieces and distributed them along the seam.

We heated the panel gradually until the solder became fluid and ran into the seam. Allow the hot metal to melt the solder. Don't heat the solder directly or it will melt and sit on top of the metal, rather than truly fill the seam.

The acid-core solder demonstration piece is a new fuel tank that Brian Zale built for our 1908 Holsman. Brian removed all of the fittings from the original tank, cleaned them thoroughly, and tinned all of the surfaces that need to seal against the tank with Eastwood's Tinning Butter. This product makes applying lead solder a breeze. The Tinning Butter acts as a flux and a tinning agent; it cleans the surface to which it is applied and lays down a thin layer of tin to which lead will happily adhere. Paint the Tinning Butter on with an acid brush, heat it until it turns dark brown, and lightly wipe it off with a cotton cloth. When applied correctly, a shiny layer of molten tin will be left behind the Tinning Butter. Brian also tinned the areas on the tank where the fittings attach. After riveting the fittings in place, Brian played a torch over the fittings and allowed acid-core solder to be pulled into the joints to seal the fittings against the tank.

When he finished soldering, he brushed soapy water on the fittings and blew shop air against their back sides to check for leaks. When the tank was deemed leak free, the soldered areas were scrubbed with a mixture of baking soda and water to neutralize the acids left by the acid-core solder.

## OLD-FASHIONED LEAD BODY SOLDER

Automotive body solder, otherwise referred to as lead, was the precursor of the plastic body filler of today. Body solder is actually composed of mostly tin and lead. The numbers you see printed on sticks or spools of solder refer to the tin and lead content specifically and always in that order—tin/lead. Nowadays there are all kinds of permutations of solder that omit lead, but for auto body repair, you will most likely see 30/70 or 20/80, meaning that the solder is 30 percent tin and 70 percent lead and so forth. Actually, traces of other metals, like antimony, are in there, but the amount is small enough to disregard for labeling purposes. The presence of tin cuts the 620 degree melting point of pure lead in half and allows melting to take place over a range, rather than at a single fixed temperature. As a result, body solder has a gloriously wide temperature range—its pasty range—in which it can be paddled about like soft butter without becoming liquid. The presence of tin also improves the adhesion of the solder to the panel. Lead does not want to attach itself to automotive sheet metal otherwise. Increasing the amount of tin makes the solder harder and less flexible, and it narrows the solder's pasty range. Therefore, 60/40 solder has a low melting point and almost no pasty range, both of which are desirable characteristics for electrical solder. As a side note, 63/37 solder is eutectic—it melts at a single temperature—so it works even better for electrical connections that need to be made instantaneously for good connectivity.

Lead is absolutely ideal for many auto body repairs because it is as tough as nails, it's flexible, it can be hammered and shaped like the rest of a panel, and it doesn't absorb moisture. Lead is not good for skim coating or covering large areas because of the amount of heat that would have to be applied, but it is the filler of choice around door and trunk openings, over factory welds along C-pillars, and anywhere else where vibration or stress would cause plastic filler to crack off in a heinous scab. We purchase our 30/70 body solder from Victory White Metals in Cleveland, Ohio, 50 pounds at a time for exactly half the cost of what you can expect to pay from one of the typical automotive restoration supply vendors.

To apply lead, assemble the following supplies: Eastwood's Tinning Butter with an acid brush, a clean cotton cloth (best) or paper towels, more lead sticks than you think you will need, maple paddles, tallow, a torch (oxy/acetylene, propane, or MAPP gas), eye protection, rubber gloves, and possibly a full-coverage respirator. Thoroughly clean the surface of all dirt, oil, rust, paint, etc. The steel needs to be silvery clean. I like 3M's cleaning discs, but Scotch-Brite discs or sandpaper will also work. If there is *any* remnant of rust, paint, or dirt, the lead will not stick. Paint on Tinning Butter in small amounts—about a 2-inch-diameter circle at a time. As noted previously, heat the Tinning Butter until it turns dark brown, and gently wipe it with a cotton cloth. Do not overheat the Tinning Butter and cook it onto the panel or you will have to sand it off. Correctly applied, Tinning Butter facilitates the creation of a bond between the body solder and the steel panel that plastic filler cannot rival. The metal should be silver and super shiny where you've tinned and wiped. Wiping too hard removes the tin, so redo any areas where you are unsure because they look dull. Always tin a larger area than you think you will need. After tinning, wipe the panel off with white vinegar to neutralize acids left

The first step in preparing to solder the Holsman gas tank was to apply Tinning Butter to all of the surfaces that would need to be soldered. We heated the solder until it turned brown and then gently wiped it, which left a layer of tin on the surface. We riveted our fittings in place and then heated our surfaces until the solder was drawn into all of the gaps. We soldered the back side of the rivets too.

over from the Tinning Butter. If you forget until after the lead has been applied, clean the metal with a metal conditioner after you have filed it. Heat the panel and scribble on it with a stick of solder as if it were a crayon. Scribble over the whole area you've tinned to lay down a thin layer of lead over the entire repair area. Now, working quickly, keep feeding in lead and paddling it out over the surface. File the lead to shape. Do not power sand because you do not want lead dust in the air. Collect your lead filings, seal them in a plastic container, and take them to a recycler. Never throw lead filings in the trash because they will eventually find their way into the groundwater.

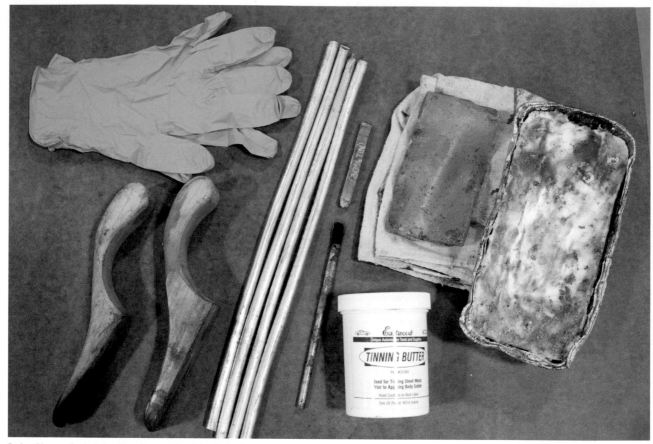

Body solder supplies include rubber gloves, solder sticks, Tinning Butter, an acid brush to apply the Tinning Butter, a cotton towel, tallow or beeswax, and a hardwood paddle or two.

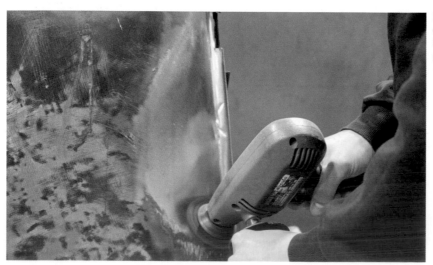

For lead solder to stick, the surface must be impeccably clean. I like 3M's cleaning discs because they conform to irregular surfaces. I dented this Ford door for educational purposes with a cold chisel.

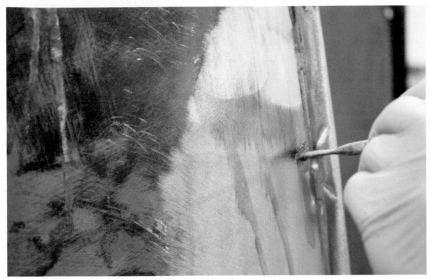

Apply the Tinning Butter with an acid brush one spot at a time. Applying too much at once inevitably leads to some Tinning Butter not getting hot enough or some getting wiped off.

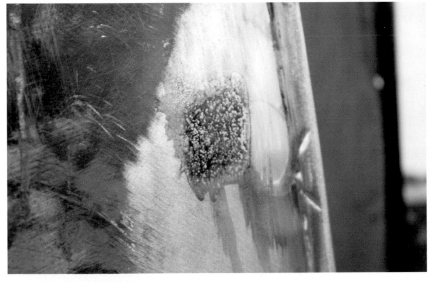

Heat the Tinning Butter until it is dark brown, but not a moment longer.

Gently wipe the hot butter once. Wipe in a direction so you aren't wiping back over areas you have just tinned. The surface should look shiny and silver. After tinning, clean the panel with white vinegar.

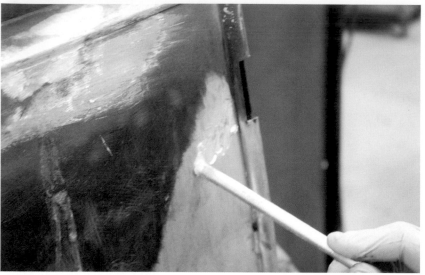

Heat the panel and attempt to scribble on it with a lead stick. Successful scribbling will leave a thin layer of lead over the entire panel. This step keeps you from adding thicker accumulations without getting everything fused to the panel.

With a little lead deposited everywhere, go ahead and pile on more lead. Heat the panel and the end of the solder stick and mush the rod onto the panel. *Mushing* is the operative verb for leading.

With more lead on the surface than you think you need, take a coarse vixen file and begin shaping the repair. Vixen files like to cut in a slightly crisscrossing action.

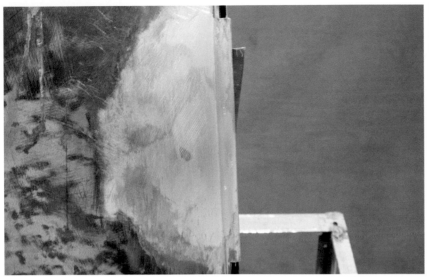

Lead solder feathers into adjoining areas beautifully. If the edges of your repair can be peeled up with a fingernail, however, this is a sign that the tinning was faulty or absent in that area.

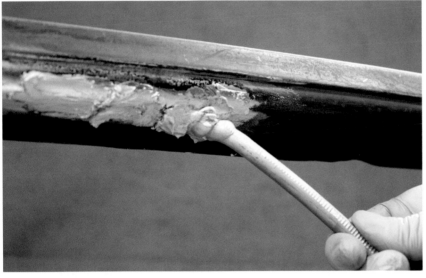

To apply lead overhead, such as along the bottom of a rocker panel, heat the solder stick about ½ inch back from where it contacts the panel. If you have a serious tremor from kicking a bad habit, all the better. Mushing the hot solder against the panel with a quivering hand is the best way to apply it.

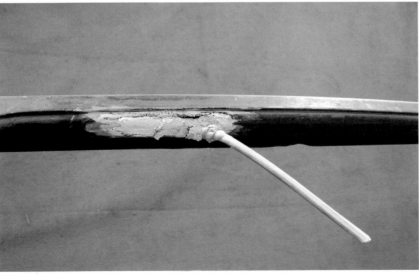

I stuck this rod upside down on a door frame to show that hot lead can defy gravity. Keep the torch moving so that the lead hovers between a solid and pasty state.

## SOME THINGS TO AVOID WHEN LEADING

Don't overwork your lead. If you get greedy and try to craft a beautiful sculpture that needs no filing or finishing, your lead will get a strange surface as its constituent parts start to separate. For the same reason, don't overheat your lead by dripping it onto the panel. Put on more lead than you think you'll need. It is difficult to go back into a lead repair. Just as with old relationships, you can't go back and recapture the magic. In the words of B. B. King, "The thrill is gone." Revisiting a lead repair that has cooled is difficult because you will have a hard time reheating everything equally as it was when the repair was first made. Furthermore, it is a good idea to wear rubber gloves any time you are dealing with lead. Lead can be absorbed through the skin, and its effects are cumulative over time. The worst things you can do are eat or smoke without washing your hands after leading. Overheating your Tinning Butter or your lead greatly increases the likelihood of noxious oxides being released into the air. Wash your hands thoroughly after you've even thought about lead or handled this page about lead. According to T. N. Cowan, who has written the excellent book *Automotive Body Solder*, wearing a respirator is not necessary while leading as long as proper procedures are followed, but it can't hurt to be overcautious.

## LEAD SLEDS

Lead is most often used in restoration to replicate factory surface treatments in areas subject to high stress or vibration, but decades ago it was used as a sculpting medium by car customizers like Bill Hines. A previous owner of our 1950 Mercury began customizing it but lost interest. The car began life as a coupe, but at some point was made into a convertible. Someone also learned to weld on it—sort of. One of the eternal truths of automotive restoration is that inherited problems from previous restorers are always worse than problems that occur naturally. The Mercury has proven the validity of that statement many times over. Anyway, it is our project now, and we are determined to see it through to completion. One area that needed attention was the sheet metal around the front turn signals, which are from a 1955 Chevy. Rather than cut out and attempt to fabricate new sheet metal to improve the transition to the light openings, we smoothed the area as customizers of the 1950s would have, with lead solder. As we added more and more lead to the low area surrounding the first turn signal opening, it occurred to us just how fitting the term *lead sled* is for an old custom such as this.

This 1950 Mercury wears 1955 Chevy turn signals. The transition between the body and the mismatched light housings is rough.

After copious amounts of lead, the Mercury turn signal on the opposite from the previous image looks much better.

## LEAD-FREE SOLDER

After completing the lead repair on our demonstration piece, a Ford Model T door, I was anxious to compare one of the modern lead-free solders with traditional body solder. I prepared a second area on the Ford Model T door and repaired it with a well-known lead-free solder, an alloy of tin, copper, and zinc. Following the instructions I found on the company's website, I applied the necessary flux, heated it until it turned brown, and then wiped it as one would do with Tinning Butter. The lead-free solder goes onto the panel very much like traditional solder, though its consistency feels waxier than lead, which is not a criticism, just an observation. The temperature working range of lead-free solder is 100 to 200 degrees higher than the range of lead, so take this into account if you anticipate working on or alongside panels without much crown. We pushed the hot lead-free solder into position using a maple paddle and allowed it to cool. Filing the lead-free solder was infinitely easier than filing lead; it was more like filing a bar of soap. Lead is remarkably dense, of course, so every time you file it you are reminded that ease of finishing is certainly one of the reasons why plastic filler came into being. Lead-free solder files more like plastic filler, but it does not feather out at its edges as well as plastic or lead, in my opinion.

Based on the grandiose claims made about this product in the sales literature—superior strength, greater safety over lead, feather-edge capabilities—and its hefty price tag, I had high expectations for its performance. Inspired by Cowan's hilarious plastic filler versus lead solder tests in his book *Automotive Body Solder*, I decided to conduct my own destructive tests to determine how well lead-free solder compares with lead. I applied lead solder and lead-free solder to two separate strips of mild steel and hammered them over in a vise. The pictures confirm lead's superior flexibility. Overall I must say that I am not impressed by my first experience with lead-free solder. It must be worked at a higher temperature than does lead, which may lead to distortion, it will not move with the panel like lead does, it does not finish out as well as lead does, and its greater safety is only relative to lead. Do not apply lead-free solder and expect to be able to sand it as you would plastic filler. Read the OSHA material safety data sheets that must be made available for these products—these sheets will not be sent with the product, but vendors will have them available through their websites. Heating and sanding these products releases metal fumes and metal dust, so do not be lulled into a false sense of security. Zinc, copper, and tin may not be as toxic as lead, but you should respect them nevertheless. Take the same precautions that you would when working with lead.

We followed similar steps to apply lead-free solder as conventional solder—tin the surface and melt the stick in place.

We filed the lead-free solder with a vixen file to bring the surface down to the correct contour.

The finished lead-free solder repair covers a series of jagged gouges that I put along the bead at the top of this door with a cold chisel.

Lead-free solder fails my destructive test of being bent at roughly 90 degrees.

Body solder that is 30 percent tin and 70 percent lead passes the destructive test.

## RIVETING

Often an automobile restorer is called upon to replicate historic manufacturing methods for the sake of authenticity. Rivets, for example, have a certain look that is hard to simulate. Although rivets have not to my knowledge been used regularly on production cars for decades, they were commonly used before World War II, so restorers of older cars should develop some familiarity with them.

Solid rivets, as opposed to pop rivets, consist of a shaft, a manufactured head, and a shop head that is formed on the opposite end of the shaft from the manufactured head when the rivet is installed. Rivets are sized according to their shaft diameter and length. Rivets are typically secured using a

rivet set, a steel tool that has a deep hole in its end for drawing the rivet down tight against the material it is intended to secure, and a shallow depression for forming the shop head after the rivet is set. The shallow depression is also handy for supporting the manufactured head while the opposite end of the rivet is being hammered. A hole through the side of the rivet set is there to evacuate small plugs created by punching rivets through very thin materials without predrilling holes. Because rivet sets typically work for only one size of rivet, you no doubt will find one day that you need to set a rivet for which you do not have the appropriate set. A lead block works quite well as a makeshift rivet set insofar as it can be used to support the manufactured head of a round head rivet while the opposite end of the rivet is being flattened. Once the lead block gets pockmarked, melt it to reestablish a clean flat surface. In a pinch, you can also drill a depression of the appropriate size in a hunk of steel or into a steel workbench top to support the manufactured head of a rivet as you set it. A homemade tool will need to be hand-held to put a shop head on, however. A few decimal drill bits near the sizes of rivets you'll use most are indispensable for fine-tuning holes and removing damaged rivets.

Try to match your rivet material to the material being riveted to avoid dissimilar metal corrosion (galvanic action). The general rule for selecting rivets is to use a rivet with a diameter equal to 2½ to 3 times the thickness of the metal through which it is driven. For riveting through two pieces of 20-gauge steel, for example, you would choose a 0.0910-inch-diameter rivet (0.036 × 2.5 = 0.090). With an automotive restoration project, hopefully you'll have some samples of the original rivets to copy. When you don't have anything to use for reference, follow the general rule and make sure the rivet looks right. Typically, old cars look quaint and, frankly, old, but not medieval. You want your project to look like an old car, not a portcullis, so avoid oversized rivets.

Apart from the appearance, the rivet size matters because the increased force required to set larger rivets easily distorts thin sheet metal around the head, so that the riveted panel ends up wavy. Undersized rivets might not have the strength you need. The hole you drill or punch for your rivets should be 0.002 or 0.003 inch larger than the rivet diameter to allow you to insert the rivets fully without undue effort, but not so large that the rivets bend sideways or fail to hold the material securely. Make sure that enough of the rivet shaft protrudes on the back side of your panel to allow you to set the rivet firmly. Shoot for an excess length roughly equal to 1½ times the rivet's shaft diameter. I will be the first to admit that this is a hassle to measure. Most people can estimate by eye and get it pretty close if they have an assortment of rivets already in hand. If you are ordering rivets, calculate the total thickness of the metal being riveted plus 1½ times the rivet shaft diameter. In real-life wrestling with old cars, you'll typically be faced with the need to trim down rivets to get the correct length. An easy way to do this is to grip the shaft of the rivet with Vise-Grip pliers situated under the manufactured head and grind away the extra material on a bench grinder or belt sander. Try to leave the new surface as flat and burr-free as

This outlet fitting for our 1908 Holsman gas tank is prepared for riveting. The proper rivet set is shown at right. We will secure the set in the vise with the cupped depression facing up. Then we will flatten out each rivet from the back side while the rivet set supports the rivet's manufactured head.

This view shows the back side of the fitting from the last image after a couple of rivets have been set. The temporary rivets (Clecos) hold everything in alignment as we go. We could use the rivet set to put a shop head on this side of the rivets if needed, but these rivets are sufficiently secured without it.

possible. Keep Vise-Grips of several sizes around for this procedure. Once you've identified the perfect width of Vise-Grip jaw for a specific project, this otherwise extraordinarily tedious job will be less painful. I usually grind several rivets, let them cool, and then quickly take off any burrs with a swipe of a file if needed. If the shaft is not ground enough and ends up too long, it will probably bend to the side when you try to set it.

Regarding the number of rivets and spacing between them, do some research on your particular project. If you have what you believe to be original panels, I would trust them more than just about any other source. Perhaps because of the level of minutiae involved, maniacal, self-proclaimed experts might lead you astray if given the opportunity to expound on rivet placement. The practical considerations are that the rivets are far enough from the edge of a panel that they can't pull out or cause the edge of the panel to flare up and that they are close enough together to secure the pieces in question. Leave yourself a good space cushion of at least ¼ inch from any edge. Space your rivets apart a distance equal to at least three times the rivet shaft diameter or they will get in the way of one another when you're trying to set them.

When riveting two things together, I find it helpful to drill each hole and secure the pieces with sheet metal screws or Cleco temporary fasteners to maintain perfect alignment at all times. If burrs are left from drilling, grind them off or the rivet will not sit flush against the metal. To set a rivet correctly, drill or punch a hole, insert the rivet, and situate the rivet set on the back side of the rivet with the hole over the end of the rivet shaft. Hammer the set to squeeze the sheet metal down against the manufactured head. With the manufactured head well supported, hammer the end of the rivet shaft to flatten it. Put the cup-shaped portion of the rivet set in place over the freshly flattened end of the rivet, and smack the rivet set to create the shop head. Steel rivets can be a bear to set by hand, especially as they increase in size. The ideal setup is to have the rivet through the material before you heat the end red hot before setting. If accessibility is an issue, it is possible, with an assistant's help, to grip the manufactured head with a pair of Vise-Grips, heat the rivet shaft until it's red hot, then push it through its intended hole, support the manufactured head against the cupped depression in a rivet set, and smack the opposite end of the rivet's shaft with a hammer. Form the shop head with a second rivet set.

If you make a mistake in setting and you need to remove a rivet, grind the head enough to create a flat spot to facilitate center punching, center punch the rivet, and drill it out with a drill bit that is slightly smaller than the rivet diameter. I prefer to grind the rivet head very thin with a pneumatic cut-off tool and pop the rivet out with a punch. You can shear off whatever is left of the head with a cold chisel instead, but that method is more likely to distort the hole and scar the panel surface.

We used aluminum rivets to attach Dzus quarter-turn fasteners to the cockpit of our miniature streamliner. There is a special tool for dimpling sheet metal to receive a Dzus fastener plate, but we improvised with a small ball-peen hammer struck over the open end of a socket with a mallet. Without the indentation, the top layer of sheet metal will not sit flush with the bottom layer despite the offset flange. Our rivets were the common rivet alloy 2117T4, which is harder than the other common 1100F rivet, which is basically pure aluminum. Either alloy would have worked fine in this case because we chose our rivet for purely cosmetic reasons from rivets we had on hand. If we need to exceed 200 miles per hour in our streamliner, at least we'll have the peace of mind knowing that we chose the tougher alloy. To learn everything you could ever want to know about rivets, consult Nick Bonacci's *Aircraft Sheet Metal*.

We set our rivets for the Dzus fasteners on our streamliner cockpit using a lead block as a rivet set. Support the manufactured head with the lead and hammer the other end of the rivet shaft to set the rivet.

Although normally the Dzus fastener plate fits on top of the panel, we put it on the underside so it would be less noticeable. We created a recess in the panel underneath to receive the plate by hitting the ball end of a ball-peen hammer over a socket. The Dzus receiver is on the underside of the bottom layer of sheet metal.

# Chapter 4
# Cutting Sheet Metal with the Oxyacetylene Torch and Plasma Cutter

**F**ortunately, the thinness that makes lighter-gauge sheet metal challenging to weld facilitates many cutting operations. Thin metal is easily cut by mechanical means, such as with tin snips, saws, cut-off wheels, shears, and various nibblers. As the thickness of the metal increases, however, the demands placed on your tools increase, and consequently, tool cost rises with increased capacity. If you enjoy being able to cut straight, curving, and freeform lines across metal as if it were paper, you need not spend a fortune, however. Luckily, the most basic oxyacetylene torch will cut carbon steels of up to ⅜ inch or more, assuming your acetylene cylinder and

torch tip are sized appropriately. An entry-level plasma cutter, though more expensive than a torch, will easily cut ferrous and non-ferrous metals up to ⅛ or ³⁄₁₆ inch. This chapter introduces you to oxyacetylene torch and plasma cutting, the processes you will most likely want to investigate to supplement mechanical cutting.

## OXYACETYLENE CUTTING TORCH

As described in the previous chapter, most metal shaping enthusiasts acquire an oxyacetylene torch because of its versatility and low cost, and let's face it—wielding a 6,000

Plasma cutters are like a lightsaber in a box. They will cut, pierce, or gouge almost any metal with ease. They are also simple to operate.

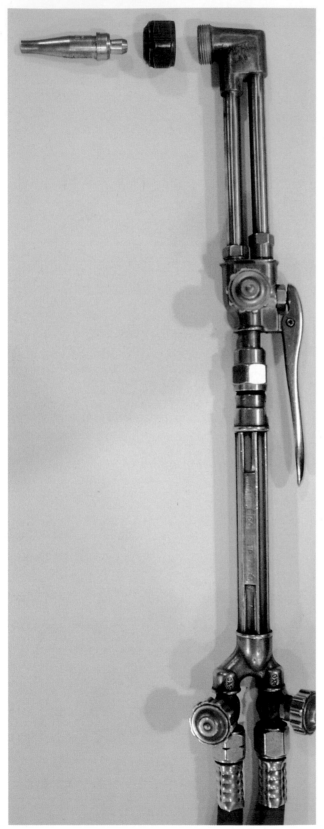

degree flame is a lot of fun. Cutting torches are available that connect directly to the oxygen and acetylene lines from the regulators on your gas cylinders, but one option is even easier. Swap the welding tip on your current torch handle for a cutting attachment. Because the attachment nut for the welding tip and the cutting attachment is tightened hand tight, the switch from welding to cutting only takes a minute. Readjust the gas pressures slightly from welding, and you are ready to cut. Furthermore, the cutting capacity of your equipment is limited only by the size of the cutting tips available for your model of torch handle and by the capacity of your acetylene cylinder.

In the welding chapter, I noted that the high flow rates of some heating, cutting, and welding tips can destabilize your acetylene cylinder by drawing out the acetone along with the acetylene. In an extreme case, the acetylene could be drawn out so quickly that it could heat up the cylinder with disastrous consequences. Remember, the customary rule of thumb is that the flow of acetylene required by your torch should not exceed $\frac{1}{7}$ of the cylinder's capacity per hour. OSHA recommendations are even more conservative, however. OSHA recommends that the flow should not exceed $\frac{1}{10}$ of the cylinder capacity for intermittent use, nor should the flow exceed $\frac{1}{15}$ of the cylinder capacity under continuous use (see http://frwebgate.access.gpo.gov/cgi-bin/getdoc.cgi?dbname=2009_register&docid=fr11au09-20.pdf). Under the more conservative recommendation, for example, a cutting tip used to cut ¼-inch steel that consumes 10 cubic feet of acetylene per hour would require an acetylene cylinder of at least 100 cubic feet.

In preparation for using the oxyacetylene torch, match the torch tip size to the thickness of the metal you intend to cut. Typically, the smallest oxyacetylene torch tip by any maker will cut any sheet metal on a car and most frames as well. Look up the fuel usage of that tip on the Internet (see, for example, http://www.airgas.com/content/details.aspx?id=7000000000128), or in the manual that came with your torch if you have it. The capacity of your acetylene cylinder might be stamped into the shoulder of the tank just below the cylinder valve. If not, contact your gas supplier to double-check the size of your cylinder. Multiply the acetylene consumption by seven or ten, depending on which recommendation you choose to follow, and determine if the resulting number is equal to or less than the cylinder's capacity. If so, you are safe to proceed.

As alluded to previously, the oxyacetylene torch cuts only carbon steel, as opposed to stainless steel and certain alloys. Torch cutting is an oxidizing process that operates on the discrepancy between the melting point of the base metal and the melting point of the oxides that form on it. For the cutting torch to work properly, the surface oxides must melt at a lower temperature than the base metal. Consequently, as carbon steel is heated, the oxides that form on its surface may be rapidly removed—oxidized—by a stream of oxygen

This Victor oxyacetylene cutting torch attachment installs on a welding torch handle in place of the welding tip. The cutting tips are selected based on the thickness of the metal to be cut.

I heated this piece of stainless steel with the oxyacetylene torch to illustrate how its surface oxides prevent the metal from being cut. The metal will melt, but the oxidizing process that takes place when you torch cut mild steel does not take place. The surface oxides, having a higher melting point than the base metal, cling to the latter during heating.

from the cutting torch. Stainless-steel, cast-iron, and non-ferrous metals typically have oxides that melt at a higher temperature than the base metal and cannot, therefore, be cut with an oxyacetylene torch. If you have overheated a piece of aluminum, you have witnessed the rubbery oxide layer that forms on the surface of the metal just before it melts. Stainless-steel, cast-iron, and non-ferrous metals may be crudely melted with a cutting torch, but they cannot be cut cleanly in the way that carbon steel can be cut.

The tip of the cutting torch has a ring of small preheating orifices surrounding a larger oxygen orifice in the center. The flame from the preheating orifices heats the metal until it is bright cherry red or hotter, at which point the user depresses a cutting oxygen lever on the torch handle. The cutting lever controls the flow of oxygen from the center orifice in the torch tip. The oxides that form on the surface of the heated metal are instantaneously transformed into a shower of sparks and molten oxidized metal by the oxygen streaming from the center of the torch tip. The resulting cut can be surprisingly clean and precise, especially on thicker metals. The cut width, or kerf, is just wide enough that you need to take it into account when cutting parts to a specific size.

In this close-up view of an oxyacetylene cutting torch tip, a ring of small preheating orifices surrounds the main oxygen orifice in the center.

To begin your cut with the oxyacetylene torch, heat the edge of your work until it turns bright cherry red, and then squeeze the cutting lever on the torch handle. The handle releases the flow of oxygen that initiates and maintains the cutting process.

Try to maintain a steady cutting speed as you move the cutting torch across the work. Note the width of the cut, or kerf. Take this into account when trimming pieces to size. If you cut directly on your drawn line, the piece may end up too small.

With the proper-size tip, correct gas pressures, and consistent cutting speed, the finished cut should be clean without an excess accumulation of dross.

Various cutting guides are commercially available to facilitate cutting circles and reproducing shapes with the oxyacetylene torch, but you may be able to replicate or fabricate cutting aids from scrap. If you will be cutting against a straightedge or guide, then you may mark your metal with a permanent marker or pencil as an aid to setting up the guide. If you plan to be able to see the line as you cut, however, use a sharpened soapstone obtained from your local welding supply.

Before you begin cutting, survey your work area for safety hazards. Torch cutting generates a lot of sparks that can travel a long way from their point of origin. Murphy's law dictates that these sparks will seek out any long-forgotten flammables that you may have stashed in a seemingly inaccessible location in your shop or garage. Even more probable, these sparks will dance unnoticed into dark crevices, where they will smolder in secret until you have

left the building. You will be halfway through dinner by the time you notice that your garage is on fire. It is a good idea, therefore, to finish any cutting well ahead of your intended completion time for each work session. In addition, keep a fire extinguisher handy and know how to use it. If most of the surfaces in your work area are concrete or steel, you may be able to do all of your cutting off the edge of a steel table. Do not try to cut directly on or near your concrete floor, however, as the trapped moisture inside will cause it to fly apart. If you have fire safety concerns, fabricate a cutting table, which is essentially a fireproof container that both supports the work and catches the sparks and molten metal created during cutting. Fifty-five-gallon steel drums are readily adaptable to this use as long as you are absolutely certain that they have not contained anything flammable or toxic previously. Also, I would not recommend a galvanized trash can either because of the potential for generating zinc fumes.

Cutting straight lines is much easier if you can find a sophisticated guide such as this length of aluminum angle. If you are an accomplished packrat or good forager, you'll find a piece of scrap that allows you to rest the torch tip at just the right distance from the work as you drag the tip along during the cut. As it turns out, this piece was a little short. An old trick you might find helpful is to install an automotive hose clamp on the torch tip to act as a spacer if needed.

Soapstone procured from your local welding supply works well for laying out designs on metal because it stands up to torch heat. Disturbing the surface of the work with a punch mark shortens your preheating time on thicker metal by creating a place for the oxidation to start.

This firebrick table works satisfactorily for cutting because it catches molten metal and keeps cut remnants from falling to the floor. Surf the Internet for examples of ingenious homemade cutting tables.

Generally speaking, do not try to cut, weld, braze, or solder old gas tanks or enclosed metal containers of unknown origin. If you do, you are asking for an explosion or at least a release of noxious gases. Do not think that you can simply fill old containers most of the way with water and be OK. Have them professionally cleaned out with a caustic solution. A gentleman who performs this service locally further advised a colleague of mine that any welding done on his freshly cleaned motorcycle gas tank needed to be done promptly after cleaning. Otherwise, he cautioned, the tank would become volatile again over time.

Outfit yourself for torch cutting as you would for torch-welding: a face shield or goggles with a No. 5 shade lens, safety glasses, cuff-less denim or heavy cotton pants and shirt, leather shoes or preferably boots, and leather gloves. Depending on your work station and the nature of the cutting you may perform, your feet are probably the parts of your body most susceptible to injury while cutting. Far removed from the action, your feet are easy to overlook in the excitement of using the torch. Therefore, keep them out of the way of falling metal and the shower of molten steel that always accompanies cutting. In extreme circumstances,

leather leggings are available that will cover your legs and boots from the knees down. Finally, take another glance around your shop. This time you are not looking for flammables as much as for things you do not want to damage, like steel rules, welding masks, painted surfaces, window glass, digital cameras, electrical cords, air hoses, and so forth.

## USING THE TORCH

Turn on the gas cylinders just as you would when welding, and set them to the recommended dynamic pressure one cylinder at a time by opening the needle valves at base of the torch handle. Because of the high flow rate of the oxygen especially, you will notice that you need to increase the pressure on the regulator a bit for the dynamic pressure to be adequate. Depress the oxygen cutting lever on the torch handle to set the oxygen pressure, or the oxygen will not flow. To light the torch, crack the acetylene valve at the torch handle base, light the gas with a flint striker, and increase the flow until the soot dissipates. Turn the oxygen valve at the torch handle base a full turn. Do not be alarmed that the appearance of the flame does not change. Now adjust the preheat adjustment knob that sits about halfway up the torch handle until the inner feather retreats back into the torch tip. This is a neutral flame for cutting purposes. Holding the torch in your dominant hand with your thumb resting on, but not squeezing, the cutting lever, place your body so that you can cut in a forehand direction while still viewing the cut in progress. The torch will be easier to control if you rest the torch handle across your other hand. Alternatively, some people prefer to pull the torch toward them; this is helpful when dragging the torch along a straightedge.

Bring the torch tip close to the work piece until the ends of the small preheat flames are about ¹⁄₁₆ inch above the surface of the work. You will know if you are too close if the tip pops. Preheat the metal until it turns at least bright cherry red. When you see a puddle start to form, squeeze the oxygen lever to commence cutting. Move the torch steadily along your cut line, maintaining a consistent distance between the torch tip and the work. If the metal is inadequately heated, the depressed oxygen lever will have a cooling effect and nothing will happen. If you move the torch along too quickly, you will outrun your puddle and the cutting will cease. Simply release the cutting lever, start preheating the area where the cutting stopped, and resume cutting once a puddle forms on the metal. If you travel too fast on thicker metal, sometimes sparks and molten metal will fly upward because you are not cutting all the way through the work as you advance. The molten metal flies up because it cannot exit the back side of the piece. If this happens, slow your travel speed down until you can see that the cut is going all the way through the metal. You will know if your travel speed is too slow because oxidized metal, called dross, will froth up in the kerf. If dross accumulates in your kerf, don't try to backtrack over your cut; backing up exacerbates dross

buildup and makes for a jagged cut. Dross will often hold the pieces together after you've made a pass all the way across the metal with the torch. If so, simply tap the cut piece with a hammer and it should fall away. Revisit any stubborn spots with the torch if necessary. You may need to cut slightly to the side of particularly bad areas.

As you cut, sparks and oxidized metal usually stream forth from the back side of the work piece a little behind where your cutting tip is. This lag is called drag and is normal on thicker steel. If you notice a large accumulation of dross on the back of the kerf, try reducing your acetylene pressure to obtain a cooler preheating flame or swap out the cutting tip for the next smaller size. Nicking the edge of a piece of plate steel, or punching its surface with a punch, will facilitate cutting at that site. When cutting thin steel of the type used for the kinds of shaping projects in this book, however, you will most likely find that the torch cuts so easily that disturbing the surface of the metal is not necessary. Likewise, you will probably not notice any appreciable drag. Instead, you may find that the torch cuts a little too well. You can cut thin metal with an oversized tip, but the cut will be more irregular and have more dross than it would with the properly sized tip. On the other hand, a tip that is too undersized for an application will never heat the metal adequately to perform a good cut. For the Victor torch illustrated in this chapter, new tips are only about $15, so you will not need to make a large investment to acquire the best tip for your purposes. For very thin metal, try angling the torch, and move it along in a forehand manner. You may find it helpful to speed up your travel speed with thin metal as well. Having the torch more upright and moving more slowly results in a wider, rougher kerf, with more dross.

Light the cutting torch with a striker, turn up the acetylene until the soot dissipates, turn the oxygen valve at the base of the torch handle one full turn, and then achieve a neutral flame by adjusting the preheating oxygen valve about halfway up the torch handle. As soon as the inner feather retreats back into the torch tip, as here, you are ready to cut.

An oversized cutting torch tip will cut like this, with a wide kerf and heavy dross buildup.

For the best-quality cuts with the oxyacetylene torch, clean the metal of any dirt, paint, or surface rust. While cleanliness is not as important for torch cutting as it is for welding or soldering, the presence of surface impurities will lead to a very jagged cut. Keep your cutting tip clean as well. Its many orifices are susceptible to clogging because the tip is spattered with molten metal from time to time. For cutting to take place, the center hole and its surrounding orifices must be able to do their job before, during, and after the cut. Any obstruction in one of these holes will degrade performance. Gently ream the orifices occasionally with a welding tip cleaner, taking care not to enlarge the holes.

To cut a hole or pierce metal with the oxyacetylene cutting torch, mark your intended target with a sharpened soapstone and preheat that spot. As soon as a puddle forms, prepare to take evasive action to move the torch tip out of the path of molten metal that might be blown back when you depress the cutting lever. Either tilt the torch sideways slightly or back the tip off from the work about ½ inch as you gently depress the cutting lever. As soon as the metal is pierced, return the torch to 90 degrees. If you are cutting a hole of a certain size, swiftly follow your drawn outline and remove the torch. The oxyacetylene torch will cut or enlarge holes beautifully, but try not to linger over the metal or dross will accumulate in the hole.

Angling the cutting tip slightly before piercing a hole helps keep molten metal out of the tip orifices.

The oxyacetylene cutting torch cuts holes with brutal honesty. The slightest quiver of your hand is cruelly manifested in the irregular shape you have just cut.

## PLASMA CUTTING

Using a plasma cutter is like tasting sugar for the first time, so beware. Some users manage to live a normal life after the experience. Others, sadly, are instantaneously and permanently addicted. The plasma cutter will cut just about any metal with speed, grace, and surgical precision. In addition, there is no lengthy setup procedure, no elaborate pre- or postcut surface cleaning ritual, and most users can learn to make satisfactory cuts through several different materials in about 20 minutes. The cost of a plasma cutter is no longer the major obstacle that it once was. As of this writing, a plasma cutter by a reputable manufacturer costs about the same as an entry-level TIG welder. Larger, more expensive machines have a greater cutting capacity and a longer duty cycle.

The plasma cutter works by creating an arc between the grounded work piece and an electrode deeply recessed in the torch handle. The air in the gap between the electrode and the work is superheated to the point that it becomes plasma, a state of matter that is neither solid, liquid, nor gas. Plasma, being electrically conductive, maintains the arc between the torch and work. The intense heat of the arc melts the metal work piece. Compressed air is fed around the electrode and through a tightly restricted opening in the torch tip to blow away the molten metal, leaving a narrow, clean kerf. Nitrogen is used in place of compressed air when the cuts must be as clean as possible.

Plasma cutters have revolutionized metal cutting. While these were once used strictly in industry because of their high cost, advances in technology and the onslaught of foreign competition has made them much more affordable.

As noted above, the plasma cutter has something for everyone. In contrast to the oxyacetylene cutting torch, the plasma cutter will cut ferrous and non-ferrous metals, and it will cut super-thin metals without distortion. Furthermore, metals can be dirty, rusty, and painted when they are cut, whereas with the torch these surfaces require additional preparation before cutting. Although you still need to be aware of fire hazards when plasma cutting, the total area that you end up heating is much smaller. This configuration makes a big difference if you are attempting to salvage sheet metal near carpet, upholstery, rubber, or glass. The plasma cutter is not as portable as a torch—you need electricity and a source of compressed air—but the machine's reach can be extended by adding longer torch and ground cables. Also, the upper capacity of the least expensive plasma cutters is going to be well under an inch, so if you need to cut really thick steel, a torch would be your best choice. For automotive applications, the small machines are fine.

To prepare yourself for cutting, follow the same fire safety precautions you would when torch cutting. For personal protection, you will most likely want to wear ear plugs or ear phones to protect your hearing. I find the hiss of the plasma cutter uncomfortably loud. If you will be looking at the back of the torch while you cut, No. 5 welding goggles will suffice. If you will be looking at the arc from the side, however, you will want something darker. I adjust my welding helmet to shade No. 9. Because you will be able to cut anything and everything, provide adequate ventilation if you will be cutting painted or otherwise coated materials. In addition, the plasma cutter carries the added danger of electric shock, so heed the numerous warning stickers on the machine. Keep the power off at all times unless you are literally poised to cut. And last, when the machine is on, treat the torch like a loaded gun. It should be pointed at the target or disarmed.

The Powermax45 used for the demonstrations has cutting and gouging capability and will cut steel up to about ⅝ inch. If you are unfamiliar with the term, gouging is the process of cutting a furrow in the surface of a thick piece of metal. Plasma cutters have a handful of small, sometimes costly consumables in the torch. The working life of these consumables varies depending on the demands made of them and on the quality of air used in cutting. Unnecessary starts, heavy use on thick material, poor user technique, and air contaminated with dirt, oil, and moisture will shorten consumable life. Make sure your shop air is filtered, therefore, to maximize the life and performance of your machine.

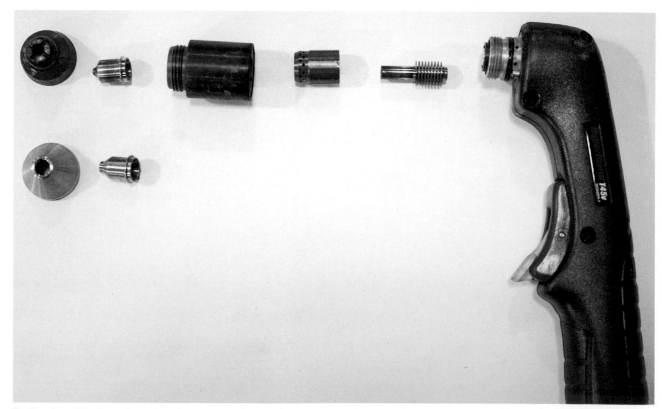

The Hypertherm 45 torch contains an electrode (near the torch), a swirl ring, a retaining cap (the large black tubular object), a nozzle, and a shield. Although the nozzle actually installs behind the retaining cap, I've grouped the nozzles in front of the cap because each nozzle fits a certain shield. The shield at top left is for cutting. The bottom shield is for gouging. The electrode, swirl ring, and retaining cap, stay the same regardless of application.

The electrode and nozzle on the right side of the photograph should be replaced. As they become worn, they work intermittently and eventually stop working altogether.

For the cutting sequences in this chapter, two different types of consumables are used depending on the type of work performed. One type of nozzle and nozzle shield are used for cutting, and another type is used for gouging. The cutting nozzle shield may be dragged across the work surface, which makes it very easy to use. The gouging nozzle, on the other hand, must held about 1/16–1/8 inch above the work surface during use.

To make a cut with the plasma cutter, make sure your consumables are properly installed in the torch. The tool illustrated has a diagram of the proper consumable orientation on top of the machine. Attach an air hose to the supply inlet on the back panel. If your machine is equipped with a cutting mode selector switch, turn it to plate cutting, as opposed to gouging or expanded metal cutting. Attach your ground clamp to your work piece near the cutting area, but not on the part that will be cut free. Power the machine on, turn the amperage knob all the way counterclockwise to the gas test position, and adjust the air pressure. The Powermax45 has an air pressure regulator screw on top of the unit and a green LED light on the front of the machine to tell you when your pressure is correct. Readjust the air pressure each time you change from plate cutting to gouging to expanded metal cutting. Each process has different air pressure requirements. Adjust the amperage by consulting a cutting chart, if available, or simply estimate where your test piece falls in relation to the machine's capacity. You can then increase or decrease the amperage once you've made a sample cut.

This view of the back of the Hypertherm 45 shows the connection for attaching your air supply. The small black knob on the top of the machine adjusts the air pressure. Readjust the pressure each time you use the machine or change cutting processes.

LEDs on the front panel of the plasma cutter alert you to problems such as inadequate air pressure, overheating, and missing or loose consumables. A green LED in the center of the pressure bar indicates that the air pressure is correct for the cutting mode selected. The three positions on the cutting mode selector switch at the bottom right corner of the panel are, from the top, expanded metal cutting, plate cutting, and gouging. The green LED at the bottom right identifies that the gouging mode has been selected.

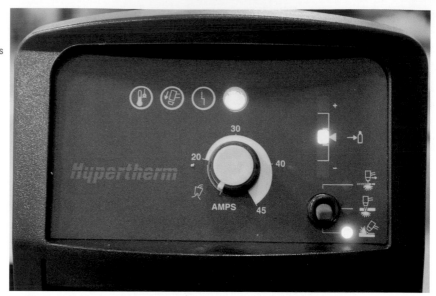

In this illustration, the air pressure needs to be readjusted, as indicated by the angry amber light above the pressure bar. Turn the amperage knob fully counterclockwise to the gas test position before adjusting the air pressure. Use the black adjustment knob on the top panel at the rear of the machine to dial in the correct setting.

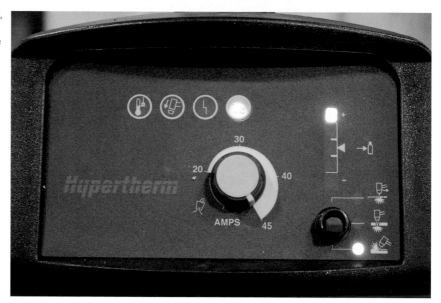

Make your first cut from an edge of the metal. Hold the torch vertically just past the edge, flip the yellow safety trigger forward and depress the main trigger. A pilot arc will start and transfer to the work piece, becoming the cutting arc. As soon as the arc penetrates the work completely, continue with the cut by pulling the torch toward you, dragging the torch tip across the surface as you go. When you are satisfied with your cut, release the trigger, let the machine hiss to cool the torch, and promptly turn off the machine when the hissing stops. You must resist the temptation to gawk in amazement with the torch pointed at your face the moment you have finished your first cut. Remember, while the machine is still on, the torch is like a loaded gun. Common sense is no match for the bewitching performance of this machine, so I am warning you now keep your wits about you if you are new to plasma cutting.

If sparks and molten metal fly up from the cut, you are either moving too fast or your metal is too thick to cut with your machine. Cutting speeds decrease pretty dramatically as work thickness increases, so try slowing down until you see sparks exiting from the back side of the work. The speed is correct when the sparks and molten metal exit at about a 15 to 30 degree angle to the underside of the work. When cutting thicker metals, as you near the end of a cut, tilt the torch slightly so that the bottom of the cut gets severed completely. A cutting guide, straightedge, or template can be used as an aid for cutting straight lines or shapes, but take into account the distance between the center of the electrode and the edge of the torch nozzle.

With your machine properly set up and your ground clamp out of the path of your cut, bring the plasma torch down on the edge of your work piece. Keep the torch at 90 degrees to the piece and squeeze the trigger. As soon as the arc has cut completely through the work, proceed with your cut.

For piercing, the plasma cutter's capacity is typically half of its cutting capacity. On thin metal, hold the torch perpendicular to the work about ⅛ inch above the surface and pull the trigger. On thicker metal, tilt the torch slightly away so that the molten metal will not be blown back at you or onto the torch nozzle shield. With the torch tip about ⅛ inch above the piece, pull the trigger. As soon as the metal is pierced, rotate the torch back upright and cut as needed.

When cutting metal to a specific size or when following a guide or template, take into account the distance from the electrode to the edge of the torch shield. This type of shield is known as a *shielded* consumable.

To cut expanded metal or metal perforated with holes or gaps with the Powermax45, set the mode selector switch to the top position and readjust the gas pressure. This setting allows the machine to reinitiate the pilot arc without retriggering after every cut. Operate the torch just as you would when cutting plate. If you forget and leave the machine in this cutting mode, it will still cut plate, but your consumables will not last as long.

For gouging, turn off the machine, swap out consumables if necessary, and restart the machine. On the Powermax45, set the mode selector switch to the lowest setting for gouging, readjust the air pressure, and roll the amperage control back clockwise to full power. Place the torch tip at a 45 degree angle with a small gap over the work and pull the trigger. Slowly move the torch along your intended path. For a shallower cut, decrease the angle of the torch. For a deeper gouge, increase the torch angle or make additional passes. Gouging can be useful when you need to butt weld two flat pieces of metal together, and they must stay absolutely flat. This seemingly difficult task is greatly simplified by thoroughly clamping the pieces to be welded to a thick metal plate with a gouged relief beneath the weld seam. Weld in small increments to avoid heating up the metal unnecessarily, but leave the clamps in place until the weld is complete. The finished piece will be as flat as the plate to which it was clamped.

When piercing with the plasma cutter, tilt the torch slightly to keep molten metal from blowing back onto the torch.

Once you have pierced the work, return the torch to the upright position and continue with your cut.

This finished cut using the circular template shows how easy it is to obtain good results with the plasma cutter.

For expanded metal cutting, choose the appropriate cutting mode using the selector switch on the front of the machine, readjust your air pressure, and proceed just as you would with sheet metal.

To gouge metal, switch your torch consumables if needed for your model machine, situate the torch at a 45 degree angle just above the work, start the arc, and proceed across the metal.

The depth of your gouge will be influenced by your torch angle. The more perpendicular the torch is to the surface, the deeper the gouge.

# Chapter 5
# Beginning Sheet Metal Shaping

Learning to form sheet metal by hand is the critical first step in your education in metal shaping. Machines are labor savers, but using them properly requires knowledge. Otherwise, they can transform perfectly good sheet metal into scrap with astonishing speed and efficiency. Fortunately, once you understand the basics of shaping sheet metal, its responses to your input will be less mysterious, so progress will come quickly—you will not need an arduous seven-year apprenticeship to start seeing results and finding satisfaction in your work. Furthermore, craftsmen have been shaping metal by hand for centuries, so do not be intimidated by the existence of complex and expensive machines, which simply harness electrical and/or hydraulic power to shape

metal according to the very same principles you will learn in this chapter. For the average person interested in repairing rust spots and making a few patch panels for a historic vehicle, for example, a few basic shaping exercises will endow most enthusiasts with the confidence to move ahead with their intended project.

One overriding principle to keep in mind when working with sheet metal is that you often trade thickness for surface area as you shape the metal. Sometimes you increase the surface area, or stretch the metal, making it longer and thinner. Other times you will decrease the surface area, often called shrinking or upsetting, making the metal shorter and thicker. I tell students to think of their metal as a slab

It's not automotive, but it's gorgeous. Ryan Brown made this handsome tail section for his Honda CB750 entirely with hand tools and the English wheel using principles you will learn in this chapter. Ryan's tail section boasts a subtle compound curve that both streamlines and enhances the otherwise chunky proportions of this motorcycle.

of dough, like a pie crust, that will react in predictable ways as you manipulate it. If you were to mash down with your thumb in the middle of a pie crust, for example, you know instinctively that the crust would get very thin under your thumb as the dough compressed. If you mashed the crust a few times in close proximity, the entire crust would spread out ever so slightly as a result. Metal doesn't behave exactly like a crust, of course, but I think this image makes it easy to understand how to change the shape of metal by influencing its thickness.

Metal can be shaped without changing its thickness as well, such as when you bend it in a vise. Think of a bend like a fold in a piece of paper; the metal is creased along a single axis. The bend could be sharp, like when you hammer a piece of metal over at 90 degrees, or the bend could be gradual, like when you bend metal around a large pipe. Perhaps at a microscopic level the thickness of the metal is influenced very slightly, but for our purposes think of the bend/fold as a change of shape that does not change the thickness of the metal. The bend or fold is easy to understand and easy to forget. Once you start shrinking and stretching, it is easy to think only in those terms, but the concept of bending is just as important as shrinking and stretching; many shapes cannot be made without bending.

I recently learned two terms from metal shaping legend Fay Butler that perfectly embody the concepts of *shaping* metal by changing its thickness and *forming* it without changing its thickness. According to Butler, *shape* and *form* were used by men such as Scott Knight and Red Tweit at the now defunct California Metal Shaping to differentiate between two distinct modes of working. Out of respect for the tradition of shaping started at California Metal Shaping, and in an effort to develop standardized terminology among metal shapers, Butler continues to use the terms. I, too, will use *shape* to refer to a process involving a thickness change and *form* to refer to a process that does not involve a thickness change.

To illustrate the idea of the relationship between the thickness and length or surface area of a piece of metal, I have taken three identical 4-inch lengths of mild steel square stock and heated two of them to make them easier to shape. I upset one by hammering on its end. The other I stretched by hammering along its side. The third was left alone for purposes of comparison. Obviously, the upset piece got shorter and thicker, whereas the opposite is true of the stretched piece. Visualize the square stock as greatly magnified versions of your sheet metal. The same changes will take place, just within a narrower plane. Keep these simple principles in mind as you begin shaping metal; they will help you achieve the results you want and hopefully answer some of your questions as you grow in your craft. Thus, whenever metal is sandwiched forcefully between a hammer and dolly or between two hammering dies in metal shaping machine, we can expect the metal to be squeezed thinner directly at the point of contact. Likewise, we can expect an

I heated and modified two out of three identical 4-inch lengths of square steel stock to illustrate the effects of upsetting and stretching. The short piece at left was upset by hammering on its end. The long piece at right was stretched, or drawn out, by hammering its sides. The middle piece is untouched. These changes take place in sheet metal as well, though they are not so easily observed as here.

increase in surface area because that squeezed metal must go somewhere—it will compress to a degree, but any metal that does not compress will squeeze out to the sides around the point of impact. There are a few exceptions to the thickness versus surface area equation, but do not be concerned with those now. Let's get a handle on the basics first.

To demonstrate the stretching effects of hammer blows on sheet metal, I will hammer a 7-inch-diameter panel of 20-gauge steel, beginning in the center and radiating out to the edges with overlapping blows.

A before and after view shows how the increased surface area brought about by hammering has manifested itself as a dramatic crown in the panel.

Let's apply the thickness versus surface area idea to create a crowned panel. For the demonstration, I have selected a hammer with a polished crowned face and a 7-inch diameter 20-gauge steel panel. Lay the panel exterior side down on a blemish-free hard surface, and work your way around the panel with light, overlapping hammer blows beginning in the center. Just like your finger in the hypothetical pie crust, the hammer mashes the panel so that it conforms to the profile of the hammer's face. As you progress from the center out, you displace a miniscule amount of the unworked metal to the outside as you go. This process trades thickness for surface area—the metal gets thinner but gains shape. By the time you've worked your way to the edge of the panel, you may have lost your mind, your elbow may never be the same, you'll have a curved panel like the one in the illustration, and most importantly, you will forever be able to predict exactly what will happen when you hammer sheet metal against a hard surface. You would get the same result by hammering a panel with a flat hammer over a rounded stake as well, only the panel would curve away from you as you progressed rather than toward you.

## STRETCHING AND SHRINKING METAL

You will be pleasantly surprised by the degree to which you can shape metal in a controlled manner simply by thinking in terms of its thickness versus its surface area. To explore this point further, draw a relaxed S curve 14 inches long onto a piece of cardboard or thick paper and cut it out. Write *stretch* on the template, as indicated in the photograph, to guide your work. Now cut a strip of mild steel or aluminum approximately 2⅛ by 14 inches. Bend a 90-degree flange ½ inch wide along its length. I am using a piece of annealed aluminum sheet in the demonstration because it is easy to form. Using a light weight cross-peen hammer or a body hammer with a narrow face, called a linear stretching hammer, lay the metal strip on an anvil or the edge of a metal table and make a series of blows along the ½-inch flange that needs to be stretched/lengthened. The small surface of the cross-peen hammerhead or linear stretching head concentrates the force of the blow in a small area, thus squishing the metal and increasing the surface area/length of that flange. Meanwhile, the rounded edges of the head leave fewer hammer marks than would be the case with a chisel pointed hammerhead, which would mar and possibly cut through the metal. Because you have lengthened the flange you have been hammering, the metal on the adjacent leg begins to curve in response to the added length. Try to match your panel to your S-curve template. If more passes are needed, change the angle at which the hammer face meets the stretched flange to prevent overthinning the metal. As soon as you feel you understand the stretching process, turn your attention to the opposite end of your test piece, which will need shrinking.

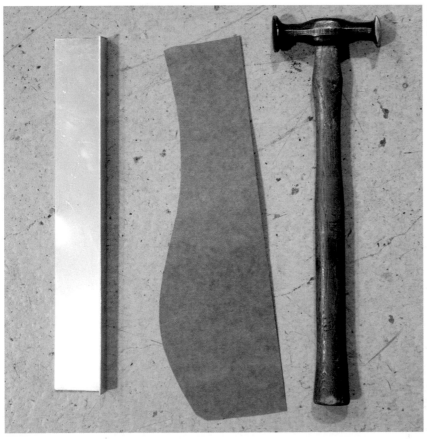

Collect the materials shown for the first stretching and shrinking demonstration. Bend a ½-inch 90 degree flange in a 14-inch piece of annealed aluminum sheet approximately 2⅛ inches wide. Make a cardboard template of a gentle S-curve, and find a cross-peen hammer with a soft radius on the cross-peen surface or modify a body hammer to have a narrow face. This aluminum is 3003-H14, 0.050 inch thick.

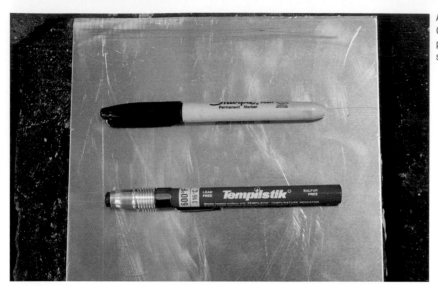

Aluminum is much easier to work if you anneal it first. Coat the surface of the aluminum with ink from a Sharpie permanent marker. You can also heat the surface and swipe it with a 600 degree Tempilstik heat crayon.

With a neutral flame on the welding tip, or with a rosebud tip if you have one, burn off the permanent ink or swipe the heat crayon across the surface. The ink burns off at about 600 degrees. Likewise, the appropriate-heat crayon will leave a greasy-looking smear on the surface at the same temperature.

As you begin to overheat the aluminum, and you will if you have not done this before, the surface will sparkle and take on a rough appearance, like sandpaper. If you keep heating past this point it will melt.

This hammer face has been ground to provide a narrow, vertical, oval contact surface. This is a linear stretching hammer, and it stretches more to the sides than at its ends because of the shape of the contact area. I have colored the noncontact area with a marker to make the footprint obvious. This concept allows you to stretch more in one direction than another. It is a recurring theme incorporated into many metal shaping tools.

Begin stretching the ½-inch flange on the work piece by hammering it along its length. As the flange is stretched and becomes longer, the piece will curve.

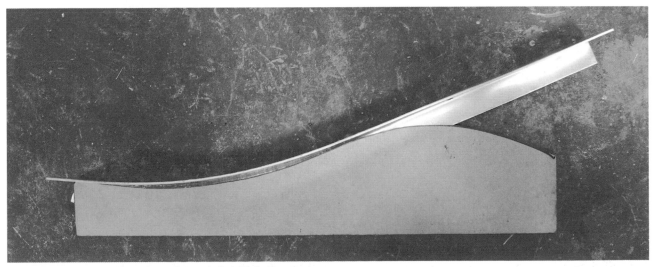

Check your progress against the template so that you don't stretch the flange too far.

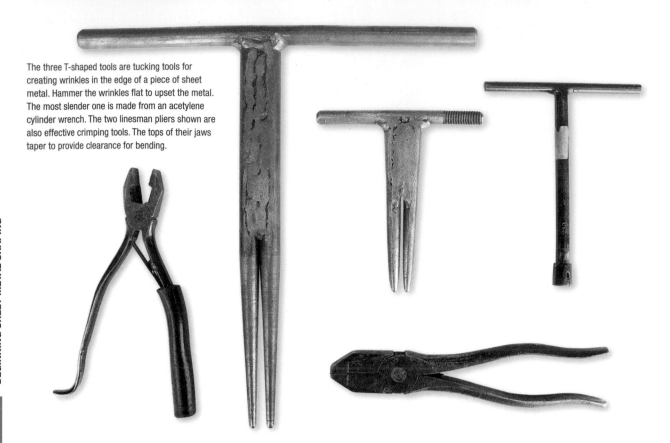

The three T-shaped tools are tucking tools for creating wrinkles in the edge of a piece of sheet metal. Hammer the wrinkles flat to upset the metal. The most slender one is made from an acetylene cylinder wrench. The two linesman pliers shown are also effective crimping tools. The tops of their jaws taper to provide clearance for bending.

Shrinking is always more difficult than stretching. One time-honored way of shrinking sheet metal is to create crimps—also called tucks or puckers—in the area needing to be shrunk and hammering the folds of metal flat, thereby upsetting the metal into itself. The best results are obtained when the tucks are restrained in such a way that they cannot simply unfold when they are struck. I will discuss this process further shortly. You can use pliers or a custom-made tucking tool to create crimps along your flange. In days gone by, metal workers also had special tucking tongs for creating puckers. These resembled blacksmith tongs, having one single jaw straddled by a double jaw.

Tucking tools may be hand-held or mounted in a vise, depending how resilient your metal is. For this exercise you want the tucks to rise up on the top side of the flange so that you can hammer them down against a flat surface. On future projects you may decide to make the tucks rise up on the back side of the panel. This decision will be based solely on which orientation gives you the most advantageous position for hammering the tucks flat. In this demonstration the crimps were easily created by hand with a tucking tool and then hammered flat against a metal surface with a rawhide mallet. When cold-shrinking, or shrinking without heat, you will be less likely to stretch the metal accidently if either your hammer or your work surface is softer than the metal work piece.

If you need more leverage than you have working freehand, you can always secure a tucking tool in a vise and create a fold by twisting the metal against the tool.

Creating tucks with pliers is easy. Just grip the flange about halfway across its width and twist to create one side of the wrinkle.

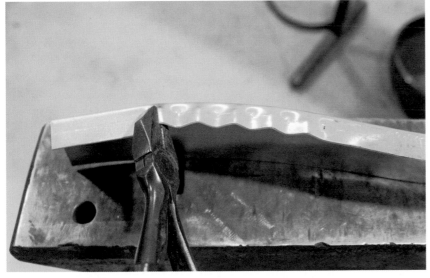

Move the pliers over to the other side of the wrinkle and twist the opposite direction.

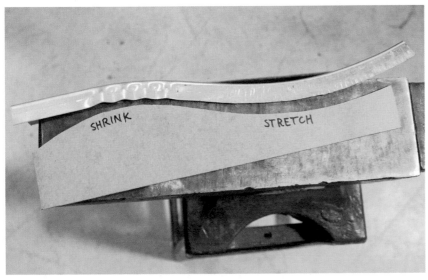

Before hammering the tucks flat, it is obvious we are moving in the right direction. The crimps along the flange have made the piece curve like the template.

On the S-curve panel, create a single crimp and first try holding the metal firmly against a hard surface by hand while you hammer the pucker flat. If the crimp wants to unfold, clamp the panel to a table on each side of the crimp so that the metal has no choice but to upset when you hammer it. If you lack a suitable table, clamp a flat steel bar or piece of angle iron to your panel straddling the pucker. If your test piece is steel, your shrinking will be facilitated by applying heat. This is called heat shrinking. Clamp the piece in one of the ways just described, heat the crest of the crimp until it's a dull red, and then gently hammer it flat with a steel hammer. The heated spot in each case will be softer than the colder surrounding metal and will readily upset or shrink. Because the metal will be soft while it is hot, you will not need to hit the metal very hard—the blow is similar to driving a tack. If you hit the metal too hard you will compress and therefore stretch it, which is the opposite of what you are trying to do. When you upset the metal by shrinking, you shorten this flange. Consequently, the adjacent leg of the panel curves toward the flange you have just shortened because of the pulling action the shrinking induces.

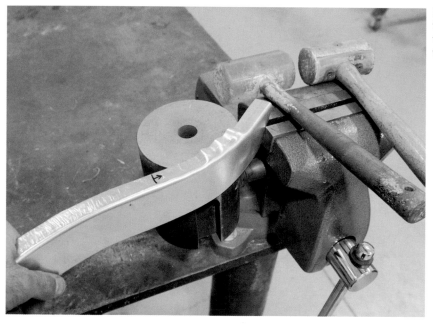

If you can keep your work piece stable as you cold-shrink, the metal has no choice but to upset. Here I have wedged the work between a stake and the vise so it will not move when I shrink the tucks.

You can secure your flange by clamping it to a table with a rounded corner. The clamps on either side of the tucks will prevent the flange from unfolding when you start hammering.

Clamping a flat bar across the curve will also prevent your flange from moving in response to your hammer blows.

Steel is more difficult to cold-shrink than is aluminum, therefore, heat may be used to soften your target temporarily. First, create a tuck or tucks with pliers or a tucking tool.

Heat the wrinkles you would like to shrink one at a time until they are red hot, and then hammer them down while the metal still has color. Hit the metal only about as hard as you would when driving a tack. Hitting too hard will compress the metal and stretch it.

It is OK to hammer the flattened tuck to smooth it, but remember that hitting too hard will stretch the metal.

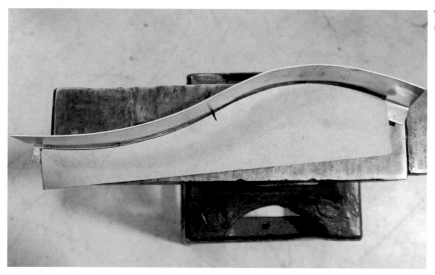

With the shrinking done, the finished S-curve panel matches the template.

The process of tuck shrinking you used on the S-curve panel is useful for shrinking metal when you do not have access to more elaborate machines for the same work. Now you will have the opportunity to explore this technique a little farther. If possible, try tuck shrinking first with annealed aluminum because this soft metal responds so well to cold-shrinking. Create a crimp in the edge of a test piece at least an inch long either by hand or using a homemade vice-mounted tucking tool. Now lay the panel down on a hard surface and begin hammering the pucker flat with a plastic hammer, starting with one hit on the outer end of the tuck, then back to the origin of the tuck, and finally working your way out to the edge. The traditional way to shrink a tuck is to start at the origin of the tuck and move toward the edge. This works fine. Ryan Heller, a former student of mine, suggested an alternative method to me two or three years ago, and I think it works better, however. Hitting the end of the tuck first creates a tiny cul-de-sac into which you can chase the rest of the tuck. By hitting the outer end first, you work-harden the end of the tuck ever so slightly so that it is less likely, in my opinion, to unfold as you hammer the rest of the tuck. Whatever sequence of hammer blows you follow, remember that you are just flattening the raised fold of the tuck to upset the metal against the resistance offered by the wrinkled sides of the tuck and the work surface. You should not hit the metal so hard that it is compressed against the table and therefore stretched. If you use a plastic hammer stretching is unlikely. Plastic hammers don't have a lot of uses, but their lack of mass and soft faces are easy on annealed aluminum. Experiment with tucks on the inside of the panel, which are shrunk against the table, and tucks on the outside of the panel, which may be shrunk against a stake. If your tucks try to unfold, try supporting the back side of your tuck against a hollowed out stump or concave depression in a piece of wood. The curvature of the wood offers additional support to prevent the tuck from unfolding.

Sometimes shrinking a tuck against a table is the easiest option. First, hit the wrinkle out on its end. This helps trap the excess metal you would like to upset.

Next, hit the origin of the wrinkle where it meets the panel.

Finally, hammer the remaining excess metal between the origin and the outer edge of the tuck.

Shrinking this steel tuck would be easier against the table than against a stake because I would have less trouble preventing the metal from unfolding in reaction to my hammer blows.

Shrinking puckers against the stake is easy with aluminum. A plastic hammer such as this one will deliver the necessary force without marring the surface.

The once proud tuck in the previous image has now been smacked into submission, but only with the amount of force required.

If you are working with steel for your tuck-shrinking exercise, you should be prepared to try the stump technique just described or heat shrink the tuck. Cold-shrinking is certainly possible with steel, but the puckers left by your tucking tool are much more likely to unfold as you work them than if they were of aluminum. Simply heat the end of the tuck until it's red hot and gently hammer it about halfway down. This method seems to be just right to prevent the tuck from unfolding, and yet it can be hammered completely flat once the rest of the pucker has been upset. Shrink the origin of the tuck and then work your way toward the outer edge of the panel.

Do not be surprised to find that tucks in steel are difficult to cold-shrink. Heat is usually necessary. First create a wrinkle with the tucking tool. In real life you would not have to clamp the piece as here thanks to the softening effects of heat, but it would be helpful.

Heat the outer edge of the tuck until it is red hot, and then hammer it over about halfway.

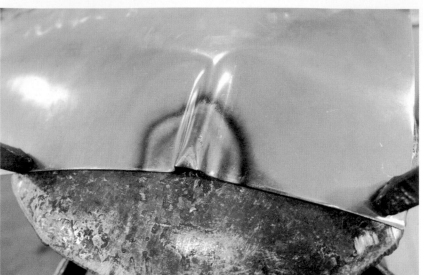

Heat and shrink the origin of the tuck where it blends into the panel.

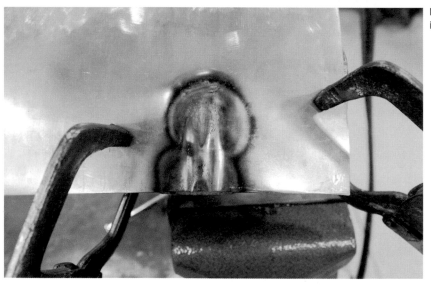

Next, heat and shrink the middle of the tuck moving from its origin out to the edge of the panel.

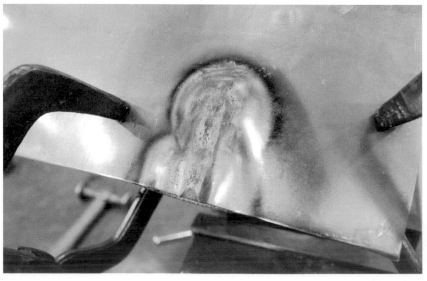

A long tuck might require two shrinks through the midsection to get it flat.

## SIMPLE MECHANICAL SHRINKER/STRETCHERS

Manual shrinker/stretchers are commonly available tools that achieve the same results you have just obtained, but through different means. Their serrated jaws grip the metal and either force it laterally to shrink or stretch, depending on the type of jaws installed. The jaws are actuated by a hand lever or a foot pedal. The foot-operated machines leave both hands free for steadying a work piece in the machine, but I find they are more easily damaged by overzealous operators. Because there is only about 1/16–1/8 inch of movement in the jaws, stomping on the pedal like a professional wrestler only damages the machine. The metal carrier for the jaws would be much improved if it were made of thicker metal. I like the hand-operated machines because you can literally feel the amount of movement taking place in the metal as you operate the handle. When operating the machine, and especially when shrinking, many smaller bites give better, smoother results than a few heavy-handed ones. The latter method tends to wrinkle the flange you're working and kinks the flange adjacent to it in the areas you've used the machine. Do most of your work on the outer 1/4 inch of the flange, rather than deep along the 90 degree bend. The metal will move more readily along the edge. If subsequent passes are needed to increase the curvature in your piece, you may move a little deeper into the flange with each pass. If your shrinking has left wrinkles in the flange, cold-shrink them with a rawhide or plastic mallet against a hard surface.

The jaws of the manual shrinker are on the left and stretcher jaws are on the right. In both cases, the wedge construction forces the serrated jaws to move the metal in the appropriate direction.

If you do most of your shrinking and stretching on the outer 1/4–1/2 inch of the flange, you will get more work done sooner and with less trauma to the piece. Notice how smooth this stretched flange is.

Proceed slowly so that you do not have to try to correct areas you have overworked. Changes of direction usually lead to unevenness along your flange.

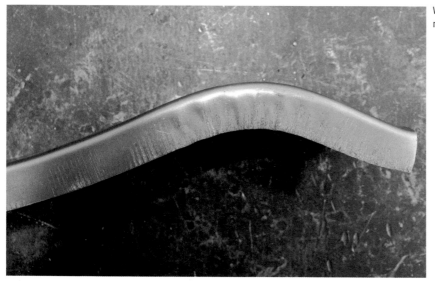

Wrinkles resulting from shrinking, like these, may be minimized through cold-shrinking with a soft hammer.

In case you were wondering how the shrinker/stretcher might be used on a car. This MGTD front splash apron has been stepped on and bent. The angle iron is being used as a straightedge.

By using the hand shrinker along the rear flange, the splash apron is made flat again.

Shrinker/stretcher machines work remarkably well and are not very expensive considering their utility. Shaping flanged pieces is their strong suit, but by no means their only use. Many times you will need a little shrinking or stretching along the edge of a crowned panel. If you decide to acquire this type of machine, I strongly recommend at least the shrinker jaws and preferably the stretcher jaws as well. As soon as possible, obtain a separate machine for each set of jaws so that you will not have to switch jaws for each round of shrinking and stretching.

## HAMMERFORMS

The hammerform is another useful tool for shaping sheet metal through controlled shrinking and stretching. A hammerform is simply a hard form over which you can shape metal by hammering. It may have multiple pieces to aid in securing the metal blank while it is hammered. Hammerforms are usually made out of wood, but aluminum is also a good choice if you anticipate needing to make multiples of something. In its most basic version, the one-piece form is shaped by cutting and sanding to match the needed part. The metal is then hammered over or hammered into the form to take its shape. The work piece is hit with a soft hammer or tool to allow cold-shrinking or stretching where needed and to avoid leaving hammer marks. Sometimes clamping is needed to keep the metal from shifting as shaping takes place. In slightly more complex hammerforms, one part of the form is cut in the shape you wish to imitate; the other part of the

form holds the sheet metal blank tightly in position during hammering. With this form, the sheet metal is sandwiched between the two halves of the hammerform and clamped to a table or held in a vise. Although close-grained hardwoods are ideal for hammerforms because they shape well and their edges don't break down, ⅝-inch-thick medium density fiberboard (MDF) works well for forms that will see limited use. MDF has no grain so it can be easily cut or sanded in any direction. It doesn't splinter, it stands up acceptably to hammering, it can be glued and screwed into thick sections, it is inexpensive, and it is soft enough to allow cold-shrinking. For my demonstration, I will use an aluminum hammerform, which is more durable than MDF, but it is overkill unless you anticipate mass-producing something or need to machine the form to obtain a finished product of exact dimensions.

To perform this practice exercise, make a hammerform by drawing a curve on a piece of MDF or wood. If you have a band saw you can screw two pieces together and cut them simultaneously. Otherwise, cut out two identical curves from two pieces of wood for the two sides of your hammerform. Sand the edge of the piece you will hammer over if it is jagged—*any* irregularities in the edge of the wood will transfer to the metal. Now cut a piece of paper about ¼ inch larger than your form along the curved portion, insert the paper between the form halves, and place the form in a vice or clamp it to a sturdy table.

The paper blank in your form will serve as a template for your metal to minimize waste. It will also clarify where

Here is an aluminum two-piece hammerform with a selection of soft hammers and corking tools. Soft hammers may be used to shape metal in hammerforms, but use discretion to preserve their faces. Hardwood or aluminum corking tools are safe choices.

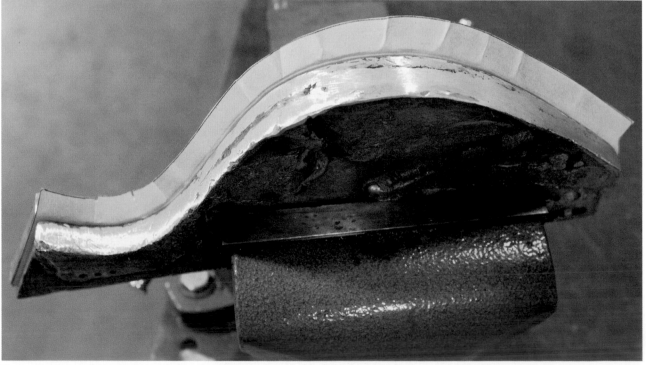

The two-piece hammerform is installed in a vise with a paper blank between the halves. Cuts in the paper indicate where stretching is needed; folds indicate shrinking is needed.

Notice that the top half of the form is slightly below the bottom half of the form to allow access the base of the intended flange. Start bending the flange over by using the corking tool out on the edge of the blank first.

shrinking and stretching will take place on your panel. In fact, the easiest way to foresee the work you will need to do on any panel is to lay a piece of lightweight paper over the intended shape, if possible, and see how the paper drapes over it. If you must create folds in the excess paper to get it to lie flat against the form, the metal will have to shrink here. Likewise, if you must cut slits or tear the paper to get it to lie flat, the metal will have to stretch there. In this hammerform exercise, the paper tells us that stretching will take place on the tight curve, while shrinking will take place on the sweeping curve. This may seem a little counterintuitive at first, but it makes sense if you imagine running a tape measure along the extreme edge of the panel before we bend a flange in it and then measuring again after we bend the flange. The metal along the extreme edge of the panel will be shorter on the large curve and longer on the small curve after the flange is added.

Now that you have seen what will need to happen to your metal, cut out a metal blank using the paper blank as a template. Adjust the halves of the form so that the top half is just slightly below the lower half of the form. This step will give you better access to bend the flange. On future projects, consider making the clamping piece of a hammerform slightly smaller than the piece you are hammering over. Now gradually begin bending over the flange with a rawhide or plastic hammer or with a steel hammer and a wood corking or caulking tool. These are simply homemade driving tools used to direct the hammer blow accurately on the blank. Soft tools do not leave hammer marks and make cold-shrinking easier. Start hammering out on the edge of the intended flange to get it to start bending. Gradually work your way along the entire flange so that progress proceeds evenly along the entire length of the piece. Do not bend a small section completely and expect to move along to an adjacent area; doing this will lead to troublesome wrinkles.

The bent flange is coming along nicely. Be careful not to let wrinkles such these on the large curve become too prominent. If they do, you can heat shrink them while the blank is in the hammerform.

113

Once the flange is bent, work your way up next to the bend itself, but do not mar the pristine edge formed by the crest of the bend. Flat spots here will stick out like a sore thumb. By working the flange slowly and evenly with the wooden corking tool, I've achieved a nice finish on the demonstration piece. In addition, take into account the thickness of the metal when cutting the size of your hammerform, especially if the piece you are forming has to fit around or over something exactly, such as an air cleaner or metal container. If you make your hammerform the exact size of the object you are copying, the new piece will not fit. A good rule of thumb for containers is to increase the size of the cover by twice the thickness of the metal.

## STUMPS

Stretching and cold-shrinking wrinkled metal comes into play with the next tool you will want to add to your repertoire—the stump. This magnificent tool has probably been used to shape metal for as long as metal has been shaped. Stump enthusiasts inevitably modify their stumps to suit their tastes, but all agree on the economy and efficacy of this old workhorse. Some people prefer a stump that stands in one place in the shop at working height, complete with tool holders attached and even a steel band that can be tightened to prevent splitting as the stump dries out over time. For people with limited space, or perhaps for someone just starting out, a smaller block of wood, such as a slice of

Thanks to the hammerform and the wood corking tool, the finished piece has a crisp, tidy flange.

One half of this MGTF rear splash apron was made with an MDF hammerform.

Our current stump has two hollowed-out areas, a crescent-shaped depression, and a concave curve (not visible). In addition, any flat surface or edge is a potential shaping or smoothing platform.

a tree trunk, makes perfect sense because it can be moved about the shop and stashed under a workbench when not in use. How big should your stump be? You will want at least a cereal bowl–size depression and a flat area a few inches wide for smoothing wrinkles. The block we use regularly has one cereal bowl–size depression, a second smaller bowl depression on one side, a dinner plate–size crescent shape cut into one side, and a concave curved side for making reverse curves. All of these cuts were made with a chainsaw, and very little time was spent cleaning up the wood afterward. The end grain of a hardwood stump would be the logical ideal choice for your stump, but I've hammered a lot of metal into all kinds of random pieces of wood, including treated landscape timbers, and it seems like anything will work for a while. Try whatever is available to you locally, and develop a feel for shaping with the stump. Then you can go about upgrading and refining your equipment. Our city's waste management facility accepts yard waste, so at any time there is pile of stumps 30 feet across free for the taking. If only I could offer a free stump with the purchase of this book.

You can bend, stretch, shrink, and smooth metal with your stump. You can also stand on it and give a speech, sit on it and have lunch while pondering your misspent youth,

or carve into it a heart with your sweetheart's name. The edges of your stump can be cleaned up for bending metal or sharpening up flanges. Stretch metal over the stump by hammering a flat sheet over one of the depressions. You need not hammer until the metal meets the bottom of the depression. Remember, your goal is to stretch the metal, *not* hammer metal to follow the contour of the stump in every case. You can stretch very subtly in this way with aluminum. Steel stretches, but not smoothly. Hammering over the depression will stretch the metal where you hammer it and wrinkle, or pucker, the surrounding metal as it is pulled toward the stretch. In this instance, the metal will behave like a napkin being pulled through a napkin ring. Tug the edge of your bedspread from a single point and you'll get a similar effect. The wrinkles or puckers that form as a result of your stretching can be shrunk by nestling them down snugly into one of the concave spaces cut into the wood and hammering down each pucker, thereby upsetting it. You can also hammer a flat sheet against a shallow dished portion of your stump to force the metal to take that contour. You can shape metal exclusively with the stump, by stretching over a hollow and shrinking the resulting folds, or you can stretch into a nearby shot bag and cold-shrink into the stump. Need the metal to shrink in one spot even though there isn't a wrinkle? Make a pucker with the tucking tool and shrink it into the stump. As you can see, few tools will take you as far as fast and at such low cost as the stump.

I strongly recommend annealed aluminum sheet for your first stump exercise because it is simple to stretch and, more importantly, easy to shrink with hand tools. For this project you will create a shape resembling one half of a motorcycle gas tank. The piece does not need to fit a real bike or even be functional. The purpose is to learn the techniques involved in shaping to build your skills.

Make several sketches on a piece of paper of your proposed tank silhouette. Don't forget to make the pattern a little larger than the finished size because the metal will be curving around the shape like a chic salad bowl. Begin by hammering the most deeply shaped portion of the tank over a hollowed out portion of the stump with overlapping

blows. As wrinkles form along the edges, hammer them into the stump to flatten them or hammer them against a stake held in a vise to shrink them. If the tank edges need to curl around more, hand tuck and shrink them or stretch the deepest part of the tank more and then shrink the folds along the edges. The crescent-shaped recess in the stump is handy in this instance because you can form against it. In about 30 minutes you will have a slightly lumpy version of one half of a motorcycle gas tank. Hammer out the lumps in the panel by laying it over a stake that matches the profile of the shape you want and hit it with a soft hammer or spoon. You can also lay the panel across a flat portion of your old friend the stump and smooth out the bumps. Define the shoulder that runs along the top edge of the tank by shaping the blank over a suitable stake with a soft hammer. I will cover metal finishing in more detail in a later chapter, but for now, this is how the tank half will look after shaping.

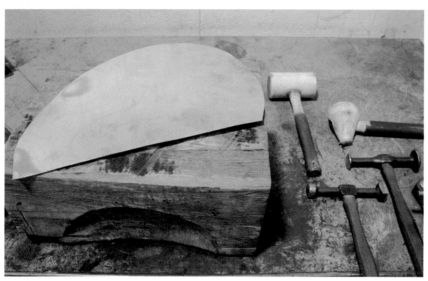

For the motorcycle gas tank demonstration, I used a plastic hammer with a round face and plastic mallet almost exclusively. You could use wood mallet or a steel hammer having one round face for shaping and a flat face for smoothing, however, as long as you are careful not to overstretch the metal. The aluminum is 3003 H-14, 0.050 inch thick.

For the first round of stretching with the plastic hammer, address the deepest areas of the part, delineated by the black marker line.

Continue stretching along the back portion of the blank. Notice the wrinkles forming around the edges of the work.

On the most prominent wrinkles, with the metal nestled into a hollow in the stump, hit the outer edge of the wrinkle first to help trap the excess metal. Then hit the wrinkle's origin, followed by the middle.

Crimp the metal with the tucking tool if needed, and then shrink it into the stump. Shrinking the edges of the blank curves them inward.

Two tucks were needed along the bottom edge of the blank to tighten it up.

The tucks along the bottom were flattened against the stake.

At this point I needed more shape in the middle of the blank, so I made two more passes with the plastic hammer in the areas marked in black.

The crescent-shaped recess was perfect for forming the bottom portion of the tank side.

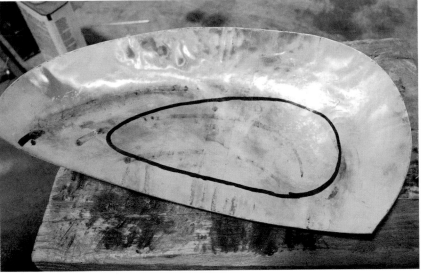

Once again, more stretching was needed inside the marked border. Sometimes it is wise to anneal an aluminum panel again during shaping to eliminate the work-hardening that has taken place. Had this area needed any more stretching after this, I would have annealed it again to prevent it from tearing.

Train your eyes to see every surface as a place on which to shape metal. I cut this stake out of a large steel pipe cap, and it works great for projects such as this one. Here I'm smoothing the bottom edge of the tank along the rear of the stake.

Once the overall shape of the tank looked right, I defined the shoulder along the top edge over a different stake.

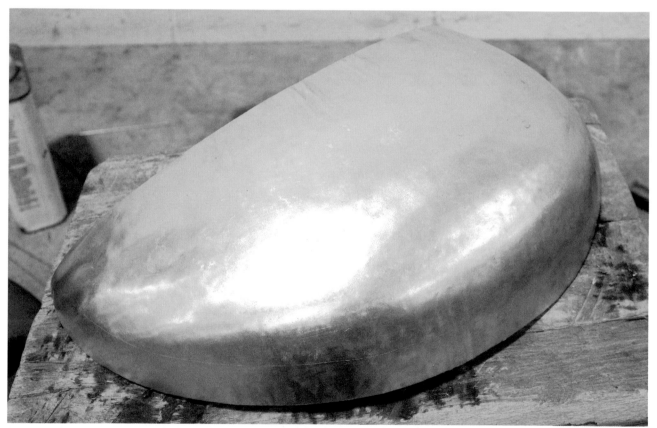

This is how the tank half looked after forming. Further finish work is needed, but it is surprisingly smooth thanks to the plastic hammers and the stump.

Now that your eyes have been opened to the lowly stump's rich possibilities, you may be less likely to underrate the next meager piece of equipment, the shot bag. At first thought, the act of hammering a piece of metal over a leather bag filled with sand or lead shot may seem primitive. Newcomers may wonder how this ritual could possibly lead to anything besides a hideous, misshapen mess. Well, after you have completed one project you will wonder no longer. Regardless of the filler material used, lead shot or sand, the bag provides a stable support for hammering, and thereby stretching, the metal into the desired shape. Shot bags and sand bags are typically lumped together whenever

Of the several shot bags we have in our shop, I think this is the best design. The top seam is reinforced, and the bag is the perfect size. The little flap visible on the side is the Velcro opening through which you can fill the bag with No. 9 lead shot. The nick above the opening is a result of a spinning sanding disc or cut-off wheel carelessly touching the bag during use.

they are discussed in the metal-shaping literature, but they are not interchangeable. In fact, they have very different characteristics. Sand is dramatically less expensive than is lead shot, but also less effective, in my opinion—unless, of course, you have a dike to repair or bullets to dodge. The sand bag is like an elusive politician; it *seems* to offer the desired support when you need it, but fails to live up to your expectations. The sand bag inevitably gives way just when you want it to hold firmly in place. The shot bag, in contrast, anticipates your every whim, providing exactly the kind of support you desire with each swing of the hammer. After only a few hammer blows into a shot bag you will marvel at how technologically advanced this homely leather bag actually is. The shot bag forces the metal to take the exact shape of your hammer's face, but only in direct proportion to the power of your swing. As a result, you can move metal rapidly or with surgical precision, if necessary, simply by changing your swing. The shot bag provides the control necessary for successful hand shaping, and it provides predictable, repeatable results.

To demonstrate the shot bag, I have chosen a deeply shaped panel from the spare tire recess of a 1929 Stutz Blackhawk. You can easily duplicate this basic shape without any guide to follow, but the process of copying an existing piece is better from an educational standpoint because there is a standard against which you can measure your progress. This shape is very similar to a portion of a motorcycle or bicycle fender, so I would encourage you to find an equivalent of those to copy. The areas that need stretching and shrinking are the same as on my demonstration piece.

Begin shaping your blank with a series of blows along the crest of what will soon be a curved panel. As you hammer through the middle of the panel, the blank will assume the general curve of the original piece along the longitudinal axis. Meanwhile, wrinkles will form along the sides that will need to be shrunk against a stake. I made a stake from a piece of steel pipe having the same radius as my panel. Check your blank against your original piece frequently to gauge your progress. You will undoubtedly find that you will need more stretching in the middle of the panel to get the original to seat firmly into your copy. Do not be surprised to find that your panel seems a little loose at its ends. This condition can be remedied by further stretching the center and/or shrinking the ends. In the demonstration I shrunk with tucks on one end and a hand-operated shrinker on the other end just to show that there is more than one way to solve this problem. When the new piece feels somewhat close to fitting the original, smooth the wrinkles and bumps out over a stake. If you are satisfied with the fit, mark and trim any excess metal from the edges. If you haven't been too heavy-handed with your hammer, as I was, your panel will look fairly smooth by now. The piece could be further finished by hand planishing with a hammer or spoon over a stake, but the fastest way to smooth it is to run it through the English wheel.

For the shot bag panel, most of the work will be done with two hammers: a large plastic mallet for smoothing and a hammer made from an acetylene cylinder cap. I have since learned that steel handles are ergonomic disasters because they transmit all of the shock from the hammer blow to your arm, so this hammer will be retired. The black panel is from the spare tire well of a 1929 Stutz Blackhawk.

Forget your self-help books, your prescription mood-enhancing drugs, and your other vices, a few blows with a big hammer and all will seem right with the world. In my zeal to create a dramatic first picture for this sequence, I have hit the panel too hard and created some wrinkles that will be visible in the finished panel. A large hammer can give you an even surface if you exercise restraint.

After one pass of stretching, big wrinkles form along the sides.

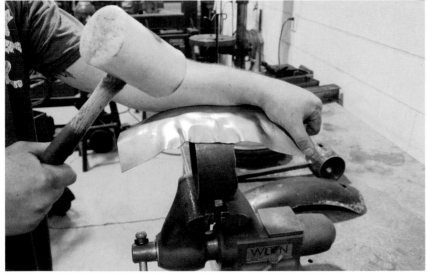

Shrink the wrinkles against a stake with a plastic mallet. Shrinking will be easier if you have a big ham hock like Matt's to sling over the panel and secure it while you hammer.

By comparing the blank on top of original panel I can assess my progress.

A second view from inside the original. I will do more stretching on my panel in the areas of the taped lines.

After completing the stretching described in the last image, the blank fits nicely on top of the original over most of the panel. The ends kick up, however.

Still more stretching is carried out along taped line to allow the original to seat more deeply into the blank. The circle identifies a low spot, and the ends are still too loose.

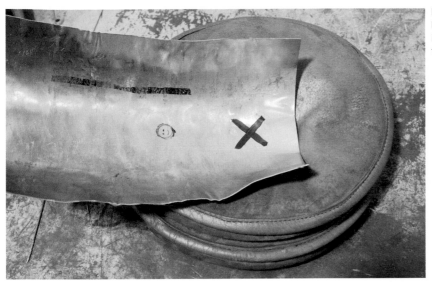

I raise the low spot mentioned in the previous picture by hammering on the X with the blank tilted up slightly in the bag to encourage wrinkles to form on the edge of the panel.

I tweak the wrinkles with a hand-held tucking tool to get them to the shape I want and shrink them against the stake.

I now smooth the entire panel over the stake to get a true reading of its fit.

The fit is much improved at the end I was just working.

At the end opposite the *O*, however, the panel is too loose.

Instead of the tucking tool I used at the other end of the panel, I used the hand-operated shrinker to tighten up this end. The moral of the story, therefore, is that there is usually more than one way to solve a metal shaping problem.

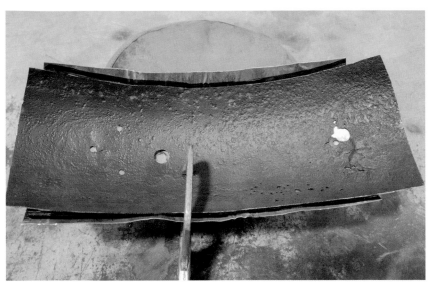

I was happy enough by now to mark and trim the excess from the edges.

The new piece fits nicely on top of the original, but it's a little lumpy.

A little wheeling across the new piece smoothes out the bumps.

Thanks to my ham-fisted hammering in the first image, a little more smoothing is needed, but the shaping is complete.

## ENGLISH WHEELS

For your first wheeling project, we will start with a 12-by-12-inch panel that you will be working from only two sides to introduce a compound curve, or crown. To create an even dome you would not wheel from strictly two directions, but limiting your choices makes the results of each pass obvious so you will learn to use the wheel much more quickly. The wheeling technique we teach is derived from the legendary John Glover. I'm sure other people have contributed to the general body of wheeling lore, but the more I learn about metal shaping the more I am convinced that Glover has contributed more than anyone else on the subject of the wheel. Watching one of his videos is time well spent (www.metalcrafttools.com). Before you begin wheeling, file the burrs off the edge of your panel and clean off any grit that may be on the panel or on the wheels themselves. Install an anvil wheel having a slight crown and make sure its contact patch with the upper wheel is as even as it can be. If necessary, shim one side of the anvil wheel axle with a scrap of sandpaper or thin cardboard.

Create a template for the intended crown for your panel by tracing a curve onto a piece of sheet metal and cutting it out. Without any pressure on the wheels, insert one edge of your panel, tighten the pressure of the anvil wheel until the panel will not slip sideways if you press on it, and push the panel through the wheels, stopping just short of having the panel exit the wheels. Pivot the panel slightly on its contact patch and pull the panel back toward you so that it passes back between the wheels about 1/8–1/4 inch from your first pass. Stop short of the edge of the panel, pivot it once more, and return across new metal to the other side adjacent to your last trip. For this pass and the next, stop about an inch closer in from the edge than on the first two passes. Stopping consistently in the same general area will result in little furrows of raised metal running parallel to the edges of the panel. Instead, by staggering your stopping points, you spread the work out over the panel and you prevent overworking the ends of each pass. It's a little like spray painting that ugly dresser in the garage—abruptly reversing your course at the edge of a painted panel results in too much paint there, making the dresser uglier than ever. Instead, it is better to carry your paint past the end of the panel. Why not take the panel all the way through the wheels then? Constantly snapping the panel in and out from the between the wheels will stretch the edges of the panel in a hurry. You could release the pressure at the end of each stroke, but doing

Note the simple, sturdy yoke beneath this anvil wheel. Depending on the design, quick-release yokes under the anvil wheel can provide an opportunity for parts to wear and create slop. This yoke has rolled miles and miles of panels and yet there is no side-to-side movement. Slop in the anvil wheel changes your contact patch and makes satisfactory results elusive. The yoke can be rotated to true it up in relation to the top wheel if needed. A narrow wheel is for tight spaces.

I would guess that this wheel has been the downfall of many eager novices. It is good for stretching specific areas and for wheeling where other anvil wheels would interfere, but it can get you into trouble. The narrow contact patch focuses the stretching like a laser. It stretches quickly, but leaves marks on the panel.

Before you begin wheeling, make sure your contact patch between the wheels is even. If the contact patch between your wheels is not even, shim the low side with a piece of sandpaper or thin cardboard. Paper gives too much.

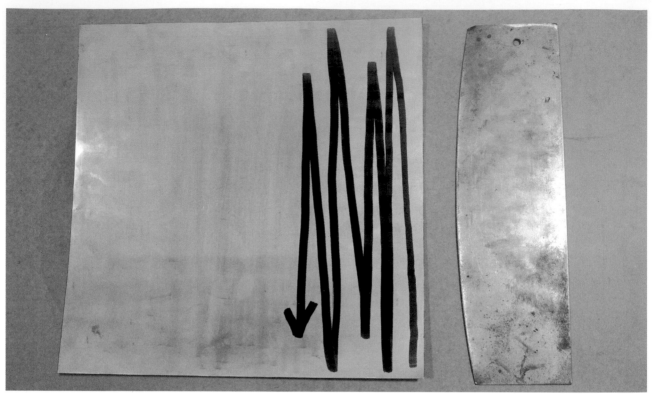

Start your first wheeling exercise with a 12-by-12-inch panel. This one is 20-gauge steel, but anything will work as long as it isn't much heavier than that. Draw and cut out a curved template like the one shown to guide you in your shaping. The black marker lines indicate the suggested wheeling pattern described in the text.

Begin wheeling at the edge of your panel. Insert your panel between the wheels and tighten the anvil wheel until the panel will not slip as you push on it from the side. In truth, you will need a little more pressure than this for most wheeling, but keep it at this setting until you get the hang of the motion involved.

Push the panel through the wheels and stop just short of the edge of the panel. Pivot the panel slightly and bring it back through the wheels over new metal adjacent to your first pass.

this would make wheeling painfully slow and it would be difficult to work the panel under a consistent pressure. Just follow the tried-and-true method of varied stops and you'll be fine. To a point, sloppiness is good in wheeling. Weaving crazily likely a drunken sailor will not work the panel evenly, however, so try to avoid that too, if possible. Don't panic, this really is not difficult. Just keep your passes close together, stop inconsistently at the edges and you'll be able to craft a smooth panel—honest.

To create a true compound curve in your panel—a curve that goes in two directions—you will need to stretch the center of the panel more than the edges. You can accomplish this by making roughly three times as many passes over the highest point at the center of the panel as you do the edges. If your first set of passes goes out to the edges, the second set might stop about 1 inch from the edge, and the third set might stop 2 inches from the edge, and so forth. Keep this overall plan in mind as you move forward.

When you get to the end of the first set of passes, release the anvil wheel pressure, remove the panel, turn it 90 degrees, and make a set of passes perpendicular to the first

set. Do not exert any upward or downward pressure on your panel as you move it back and forth through the wheels. No bending is involved in this exercise *at this time*. Check your panel's fit against the template as shown. You may not notice much change because not much pressure is on the wheels. Repeat the wheeling process from both directions with more pressure. The tracking pattern will be discernible on the surface of your panel as you progress, but if the individual passes have distinct edges, giving the piece a faceted or fluted look, back off the pressure exerted by the anvil wheel. When this happens, and at some point it will, reduce the pressure and make a general pass over the panel from the opposite direction to smooth the surface. Fortunately, one or more passes over the entire panel under light pressure from two directions will cure most panel ills by working out uneven stresses. Use this method as your fall back plan any time you run into trouble with the wheel.

As you experiment with the English wheel, keep in mind that it is foremost a stretching machine. With the flatter anvil wheels installed, the English wheel creates an arched shape that predominantly curves in the direction of

**131**

It may take a few minutes to master the art of steering your panel, but eventually you will get to the end of your first set of passes. Read the text describing how to vary the length of your stroke through the wheels. Release the pressure on the anvil wheel, remove your piece, rotate it 90 degrees, and repeat the wheeling process across the panel from this direction.

the rolling wheels because the contact patch is an elongated oval. With highly crowned anvil wheels, the contact patch is almost round, so the wheels stretch in 360 degrees; you will get more wheel marks in the panel and the pattern of stretching will extend in all directions. To state this in a different way, if you are standing at the side of the wheel like Glenn in the illustrations, the panel will curve in line with your outstretched arms when the flatter anvil wheels are installed. Therefore, whenever you consider sending a piece of sheet metal through the wheel with even moderate pressure, be prepared to create a hump in your metal arching in the direction of your outstretched arms. I like the term *hump* because it implies more shape in one direction than in another—think of a camel—which is exactly what happens with the wheel. The panel will curve upward in the direction of your outstretched arms more than it will across the panel the other way most of the time. Let's see how these ideas apply on some sample panels.

Hopefully, after you've made a few passes through the wheel, your panel will start to curve like the one in the illustration of the first sample panel. The template, when laid

across the middle of the underside of the panel, touches only in the middle of the panel, but it touches more of the panel than it did when the panel was flat. When first learning to use the wheel, you will be tempted in situations like this to see the part of your panel that touches the template as *correct* and think that the edges must be *wrong*. Your inner monologue will go something like this: "This middle part touches the template, which is the goal, so it must be right. I need to do something to the wrong parts to make them right." There may or may not be expletives involved, but don't lose heart; you've only forgotten that your sole option with the wheel is additional stretching. The solution, therefore, is to keep wheeling the center part of the panel to add more shape there; raise it up through more stretching so that the nonfitting panel edges will fit.

Another common wheeling problem you may encounter in this exercise is unevenness in your panel, as seen in the second sample panel illustration. The demonstration piece touches the template along two strips, one on each side of the centerline. The unevenness results from uneven work on the panel. Too much stretching has taken place in the areas that rise above the template. The other way to look at it, of course, is that the low spots—where the template touches—haven't been wheeled enough. The solution is to wheel the low spots at 90 degrees to the orientation of the template. After spot wheeling to remedy situations like this, it is helpful to make one or more passes over the whole piece from two directions under light pressure to even out the panel.

By now you are probably really enjoying yourself. Wheeling is fun and therapeutic. In fact, the *whoosh-whoosh* of that siren's song may have caused you to overwheel your panel so that only the edges touch the template, as in the third sample panel illustration. Now that you know your only option with the wheel is additional stretching, carefully wheel those edges, easing off the pressure as you blend the new stretched areas into the first couple of inches along the edges of the panel. Be extra careful when wheeling edges, however, as they are very easy to overstretch! If they become overstretched, the edge of your panel will look like an eyelid, as the excess surface area manifests itself as a hump. This situation is psychologically disturbing and a hassle to get rid of. Run the edge of the panel through the hand-operated shrinker to tighten it back up, and then smooth out the marks left by the shrinker on the wheel. If you are able to overcome the aforementioned difficulties, your panel will fit the template across its middle from any direction.

If you lay the template across the middle of your panel and it looks like this, you do not have enough crown. Keep wheeling to stretch the center. Remember, the template will only fit across the middle of panel, but it will do so from two directions—north/south and east/west—if your panel is even. Remember, too, that the wheel adds more shape in the direction of the rolling wheels than the other way with the flatter anvil wheels. Hold your template perpendicular to your tracking pattern when checking your progress.

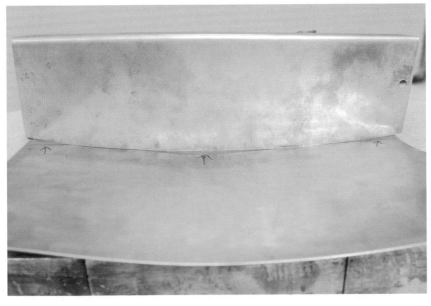

If your panel looks like this, you have worked the panel unevenly. Make some passes perpendicular to the template in the areas that currently touch the template to stretch them. Raising these areas will eventually make the template fit across the panel without gaps.

Whoopsie. If your panel looks like this you have wheeled too much and put too much crown in the panel. Carefully wheel the extreme outer edges, and then ease off the pressure slightly on the anvil wheel to blend the newly stretched edges into the rest of the panel.

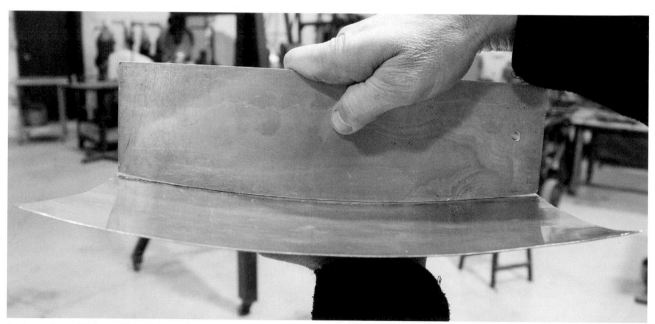

This is a pretty good fit. I see a little light over on the right edge, but I'm not going to mention it to my volunteer, Glenn.

Now that you understand how the English wheel works the metal, take the panel you've just created and see if you can create a true dome so that your template will fit across the diagonals as well as across the sides, as in the first part of the exercise. It may not take much work if your first panel is even. The only trick here will be keeping the very center of the panel from becoming overworked and therefore picking up too much crown. Be sloppy with your passes and avoid travelling directly over the center of the panel except very rarely. Roll across the diagonal areas that seem a little low and turn the panel frequently to blend the previously unworked areas into the rest of the panel. When your template fits your panel 360 degrees, you are ready to tackle any simple crowned panel.

As you may have guessed, this is only just the beginning of wheeling. Now we'll look at some other useful techniques. Not every panel you will want to create needs a crown or compound curve put into it; some are simply bent on a gentle curve. If this is the case, you can use the wheel as a kind of rolling sheet metal brake to induce the curve you desire. Install the flattest or second-flattest anvil wheel, insert the panel, and increase the pressure between the wheels just to the point where you can push the panel sideways and it won't slip. Pull the panel toward you and gently pull down at the point you wish for the curve to start. At the end of the stroke, jog the panel over slightly just as you did in the first wheeling exercise, and push the panel back through the wheels with as little resistance as possible. Do not try to bend on the push stroke or the panel will inevitably bend prematurely. Repeat this process until you have carried the bend the full width of the panel. If you find that some parts of the bend need adjustment, simply repeat the process and either pull down more sharply to increase the bend or lift up as you pull the panel toward you to decrease the bend.

The technique just described may be used to introduce a bend in a crowned panel as well, either to increase the curvature abruptly or create a longitudinal bend, but there are some caveats. To increase the curvature of a panel, use light pressure and simply pull down on the panel as you pull it toward you through the wheels. If the pressure between the wheels is too great, the stretching will increase and a hump will form. For a longitudinal bend, select an anvil wheel with enough clearance so that the edges of the wheel do not interfere with the panel and create unwanted ridges in the work piece. Run the panel lengthwise and apply gentle downward pressure to the panel alongside the wheels with your hand. Be mindful of the effects your bending have on the panel though. If you try to bend along a curve, as in the demonstration, what are you asking the existing amount of metal to do after the bend? Take up more space or less? This is really just another version of the large curve on the hammerform panel exercise earlier in the chapter. If you are trying to create a bend along the outside of a curve as here, where the distance covered by the metal will be shorter after it is bent, shrinking will be needed. If the short side is near an edge, you may be able to use the hand shrinker or tucking tool to get the results you want, but if the curve is deeper in the panel you will have trouble shrinking the metal by hand, though there are machines capable of doing it. Instead, put the needed curve into the panel early in the process, and rely on stretching to bring out the shape you need, or make the panel out of more than one piece using smaller sections that need only stretching.

To bend a gentle curve in a panel using the wheel, install the flattest anvil wheel without sharp edges, use light pressure, and pull down gently on your panel as you pull it toward you on the area where you hope to produce a bend.

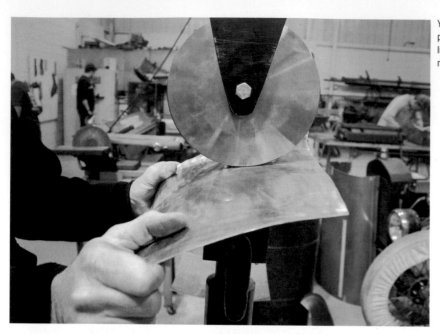

You can increase the curvature of a shorter panel, too, by pulling down as you pull the panel through the wheel. Use light pressure and be gentle or you will overstretch the metal and possibly leave wheel marks.

It is also possible to create a bend longitudinally by pushing down on your panel next to the anvil wheel, but be aware of the effects your bend may have on the rest of the panel. If Glenn were bending more than a slight crease on this curve, some shrinking would be needed below the apex of the curve. After all, he would be attempting to fit a certain length of metal into a shorter space below the curve.

The English wheel can also function like a bead roller for introducing delicate details into the surface of panels. For example, if you need to reproduce a crisp line in an automotive body panel, install the flattest anvil wheel and adjust the pressure until the panel will not slip sideways when you push on it. Slowly work the metal back and forth along the needed bend line while pushing down on the panel alongside the wheels. For additional crispness, you can add a thin shim under the pinching side of the anvil wheel's axle. Many automotive panels have subtle creases or body lines that are prominent in one area, but fade into the panel at the opposite end. In the demonstration, Ryan pushes up on the panel while working it back and forth through the wheels to create such a body line. Gradual progress is the key here. Don't just jam the panel in there and expect to bend a clean line with one fell swoop. Creep the panel backward and forward while lifting up on it alongside the wheels. After creating the bend you need, you can wheel or hammer the new segment to add shape. In the demonstration, Ryan hammers the edge of the panel to create a curved addition that fades into the original portion of the panel.

Use the wheel like a bead roller to create crisp edges in panels. Install the flattest wheel and use light pressure.

It is possible to roll a crisp line all at once, but it is easier to sneak up on it incrementally, a little like sketching a circle with a pencil, just use light pressure on the wheels and only push down on the outside of the panel when you've found the perfect line. Depending on the edge radius of your upper wheel, you can bend against it as well.

Similar, but more drastic than the example in the last photograph, you can use the wheel to introduce significant lines or directional changes into a panel.

Roll the work piece back and forth while pushing up alongside the wheels to establish the bend line a little at a time.

Use light pressure on the anvil wheel to blend the newly defined area into the rest of the panel.

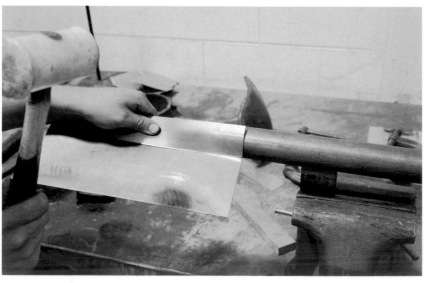

This picture sequence may give you some ideas for building upon these wheeling techniques. Here Ryan takes the panel from the previous picture and adds curvature to the newly segregated section of the panel by hammering it over a stake.

More shape is added along length of the panel with the hammer, but care is taken to blend the curvature in with the rest of the panel.

This view of the panel upside down clarifies the work done in this example.

## BEADING

If your bead roller has a reliable depth gauge, you can adjust it to the desired distance from the intended bead and use the gauge to guide you as you roll a bead. If you don't have a depth gauge, or if your panel has too much shape to use the gauge, you will have to roll the panel freehand. To prepare your metal for receiving a bead in the bead roller, paint the surface with a machinist's dye, such as Dykem, and scratch away the intended bead path with a pair of dividers. Beware that a true scored line can lead to a crack, so just skim the surface

enough to remove the dye. A permanent marker works well too, though it is harder to maintain a consistent bead width along a curve without additional drawing instruments, such as a French curve. Always run a test piece through the machine first to make sure it is set up correctly, and then run your panel through at a slow steady rate using two or three passes, slightly increasing the pressure each time. Unless you have orangutan arms such as mine, you might enlist the help of a friend for this operation. Both the cranking of the dies and the guiding of the work piece are critical to rolling a bead successfully.

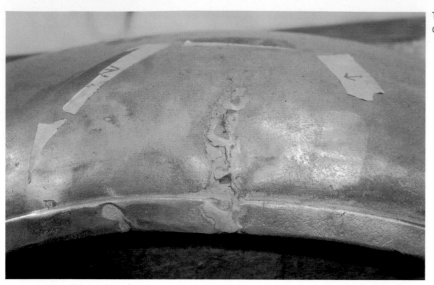

This old ghastly repair of a beaded steel fender is typical of what old car folk can expect to find on their projects.

Fortunately, Michael Spyropoulos was up to the task of repairing a similar catastrophe as in the last image on this fender.

Each time you go to roll a bead, make sure the dies are aligned, unlike these. Keep in mind, however, that changing the alignment might make all kinds of new beads possible.

One good way to mark metal for beading is with machinist's dye and a scribe or set of dividers. Spray-on dye is faster than the paint-on kind, but the latter shows up well in photographs.

Mark the back side of your metal, adjust the machine on a test piece, and roll your bead. You will probably want someone to crank the machine while you guide the work piece through the dies.

If you are fortunate enough to have a machining background or know someone who does, your beading options are limited only by your imagination or your friend's generosity. Nevertheless, even if neither of these things is true, affordable outfits are available to equip you with a machine and a collection of dies. One set of dies we use frequently at McPherson College is the homemade equivalent of commonly available V-grooving dies. The top die has a knife-edge that rides in a grooved lower plastic die.

These dies are handy for creating subtle creases and fold lines for flanges. Another option for creating flanges is a set of burring dies. For decades these were standard equipment in sheet metal shops where metal containers, metal products, and air ducts were made, so you might be able to find an old bead roller with these dies at a local auction, at a flea market, or on the Internet. A narrow rounded-top die also works well for rolling a flange when used with a smooth bottom die.

Another widely used bead in automotive and other applications is the wrapped wire edge, which bolstered the rigidity of sheet metal products ranging from watering cans to luxury cars. In the event you find an old beading machine, it is possible that it will have a set of turning and wiring dies to go with the burring dies just mentioned. These dies, too, were very common for years. Every old sheet metal book I've seen describes the process for creating a wire edge on a panel

using the beader, but I have never been as happy with the results I've obtained as when I do it by hand, probably because automotive sheet metal is thicker than the sheet metal used in ductwork, drain pipes, and sundry other metal applications covered in metal shop books of yesteryear.

To create a wire-edged panel by hand, bend a soft 90 degree flange along the edge of your panel by any means at your disposal—bend it in a brake, bend it in the English

These dies work well for creating flanges along the edge of panels and for creasing panels.

By increasing the pressure on dies as you progress, you can roll a bead that disappears into the panel. Start at the faint end and increase the pressure as you go.

These burring dies in a Pexto bead roller work well for bending flanges. Adjust the gauge adjacent to the wheels to the correct width for your flange and start rolling with very little pressure exerted by the upper wheel. Tighten the pressure a little, and then make several passes, lifting up on the outside of the panel each time.

This narrow rounded-top roller with a flat bottom roller works great for establishing a bend line for a flange. Make a flange like this first if you want to create a wire edge on a panel.

The back side of this automobile fender bead gives you a good view of a wire edged panel. The tightness of the crimping around the wire varies from one example to another.

wheel, use a bead roller, or Vise-Grips. I like to use the beader with a flat-bottom roller and a thin rounded-top roller. The flange should be exactly 2½ times the diameter of the wire you will wrap. Hammer the top edge of the 90 degree flange over a spoon or dolly with a ground edge to create a U-shaped channel. The U-shaped channel is essential. Trap one end of the wire in the flange with a hammer, a pair of Vise-Grips, or a homemade tool such as the one I have illustrated. Gradually work your way down the wire, crimping the flange around it as you go. If you make a wiring tool such as the one I've shown, make the first pass with the concave jaw squeezing the flange to trap the wire. Flip the tool over for the second pass and tightly crimp the wire with the flat jaw of the tool. This tool makes perfect wire edges easily compared with any other hand method I've seen. Without the special tool, wiring by hand will seem tricky the first time you try it, but it can be done with just a hammer with practice.

After you bend a 90 degree flange along the edge of your panel, hammer it over one of the objects shown to create a U-shaped channel to receive the wire. Heed the following suggestions: slide the hammer along the table surface at a slight angle so that it hits the flange high; for curves, use a curved face hammer so you won't leave hammer marks; bending the U will be easier if you grind a tool to fit into the U-shaped channel perfectly.

Hammer over the open channel you've just created by hand to trap the wire or use a homemade wiring tool such as the one in the next image. You can also modify a set of Vise-Grips to bend the channel over a spacer welded to one of the jaws, but you have to be willing to sacrifice a set of Vise-Grips.

This highly sophisticated crimping tool, made from an old pair of nippers, clamps down on the loose edge of the flange and traps the wire for the first pass.

For the second pass down the panel, flip the tool over and tightly crimp the flange around the wire with the flat jaw. Do not roll the tool over the wire under pressure or you will scar the panel. Crimp as needed, release the pressure, move the tool, and crimp again.

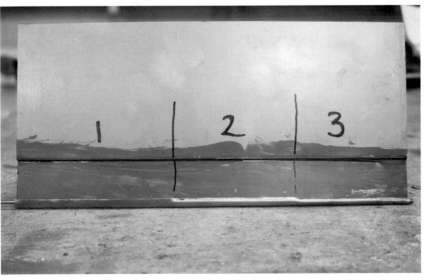

This panel shows the three steps involved in creating a wire edge: 1) bend a flange into a U-shape; 2) trap the wire with the wiring tool; and 3) crimp the wire tight.

## COPYING SHAPES

Now that you have added several metal-shaping techniques to your repertoire, I want to describe some ways to copy or create shapes upon which you can apply your new skills. For existing objects that you want to replicate, like straight but rusty fenders, you may not need a buck or rigid three-dimensional model. If you can lay the piece you are making over the original and get a satisfactory indication of the accuracy of your panel, there is no need to make extra work for yourself by creating a buck. Often, however, overlaying an old sheet metal panel with a new one is deceiving. The new panel might fit over its original source panel beautifully at the edges, and yet there might be a discrepancy somewhere in the body of the panel. You often need some additional guidance in the form of one or more templates of various contours to get your copy spot-on with the original.

A profile gauge like carpenters use to copy wood trim molding is probably the easiest tool for copying a shape. Mark off template locations on the original piece with a permanent marker, press the profile gauge onto the panel, and trace the resulting contours onto stiff cardboard or sheet metal. Index and label each contour with other templates that it crosses for maximum accuracy. Areas with a lot of shape change require more templates. As you begin shaping your flat sheet metal, the templates will tell you what you need to do to your panel. If your templates represent the outside contour of the desired shape, contact between the outside edges of the templates and your panel means that the center must be stretched, for example. This is just like the first wheeling exercise on the crowned panel in this chapter except that the template fits the outside of the panel rather than the inside. As you shape your piece, check it against the original as well

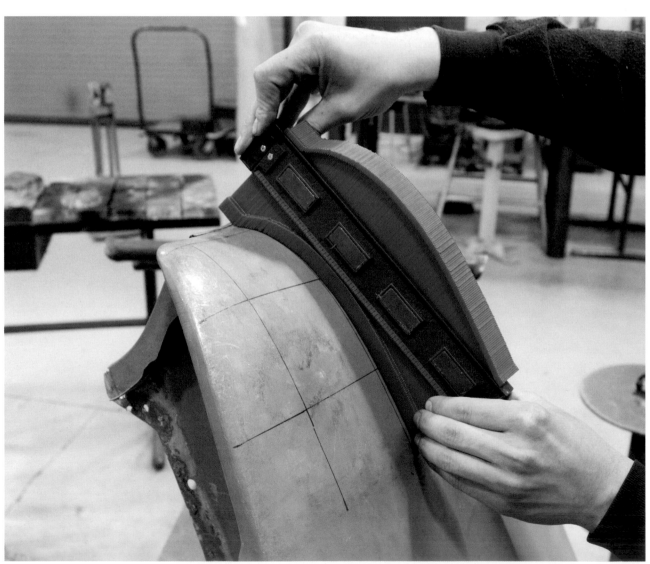

A carpenter's profile gauge allows you to record shapes quickly. Trace the curve onto a piece of stiff cardboard and cut it out, and you've got a great template.

For longer spans than a profile gauge can accommodate, shrink and stretch the edges of a strip of steel until it matches the contour you'd like to copy.

as the templates. When your panel starts to get really close to fitting, lay it over the original, if possible, and scrub it gently to and fro. The friction will leave small scratches on the inside of your panel that indicate where more stretching is needed. Circle those areas with a permanent marker and continue working the panel.

If you don't have a profile gauge, another method you can use to copy a shape is to shrink or stretch a strip of metal with the hand shrinker/stretcher. You now know that stretching adds length and curves a piece of metal away from you, while shrinking shortens the piece and pulls it toward you. Take a strip of sheet metal about an inch wide and as long as the contour you need to copy and work the outer ½ inch along either side with the shrinker/stretcher until the curve of your piece matches the contour.

You can also copy shapes by tracing them with a compass. Place a piece of cardboard vertically along the contour you wish to copy. Beginning at one end of the contour, adjust your compass to the largest gap between the panel and the cardboard, and drag the compass along the shape, holding it perfectly vertical the entire way. Cut out the excess cardboard and trace the shape again. After the second pass, the traced contour will be correct. If you tilt the compass to the side, you'll end up tracing a contour slightly larger than the original. You can also trace a contour using a set of dividers to scratch the contour onto sheet metal, but the latter method gets awkward on large pieces.

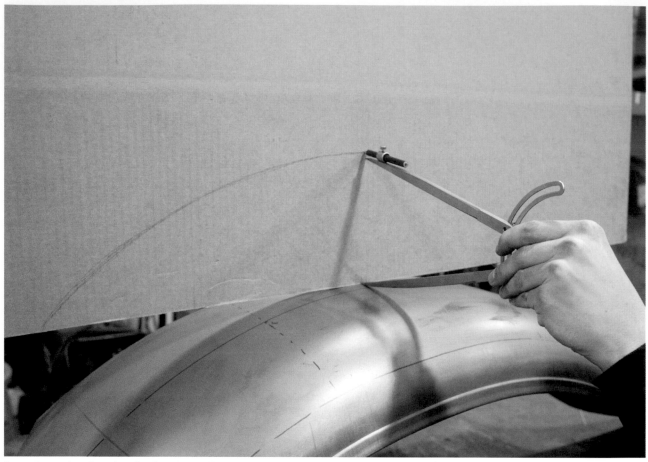

For very long contours, lay a piece of cardboard on top of your shape, adjust a compass to the largest gap between the work and the cardboard, and trace the contour. Keep the compass perpendicular the whole time. Trace the shape once, cut off the extra cardboard, and then readjust the compass and trace it again for a perfect match.

Another method of copying shapes relies on flexible patterns. This process was described to me by metal shaper Wray Schelin, who assisted in its development with artist Jonathan Clowes and his assistant Mark Goodenough. These gentlemen were discussing ways to cover a conduit frame for one of Clowes's sculptures with a flexible pattern in hopes of facilitating the fabrication of the sculpture's aluminum skin. Goodenough suggested that they use a special sign maker's tape, called transfer tape, to make the pattern easy to remove from the conduit framework. Transfer tape is tough, flexible, and not overly sticky, but to be usable as a pattern it needs to be more rigid, Schelin explained. Schelin suggested that a double-layer overlay of fiberglass-reinforced shipping tape be applied to the back side of the pattern before peeling it from the subject. After peeling the pattern free, it was dusted with baby powder to make it less sticky. In the end, the pattern worked perfectly for the sculpture.

Flexible patterns may make rigid templates and wood bucks obsolete for some craftspeople. I will demonstrate a buck in more detail in the next chapter, but I can tell you they take a long time to make and they take up a lot of room, even

if they are disassembled. Furthermore, wood bucks become even more complicated if you need to make two mirror-image versions of something. There are contour changes that preclude just flipping all the buck's stations 180 degrees to create a mirror image.

We tried the flexible pattern technique on a Plymouth Belvedere rear quarter panel, and our pattern came out exactly as Schelin described. It accurately represents all of the contours of the panel. The only drawback I felt was that there was still some ambiguity regarding where different contours relate to one another in space. Perhaps the flexible pattern, when used in conjunction with a few firm templates, will prove to be the best way to copy shapes after all. The flexible pattern seems to me to capture the correct shape, achieved through thickness change, but not the form. Schelin plans to make a DVD about this process in the near future, so perhaps some secrets are soon to be revealed. In addition, you will not want to pay for a big roll of transfer tape just to try out the technique. Visit your local sign maker and see if someone working there has a remnant he or she would be willing to give you.

To make a flexible pattern you need transfer tape, a small plastic squeegee and some fiberglass-reinforced packing tape. Sprinkle some baby powder on the finished pattern so that it won't be sticky. This transfer tape is from American Biltrite (www.abitape.com).

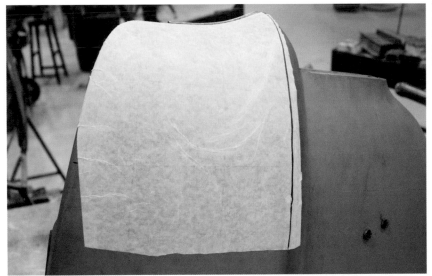

The first step with the transfer tape process is to cover your shape evenly. You may have to cut the tape for abrupt shape changes without creating wrinkles. Push out any air bubbles with a plastic squeegee or plastic body filler spreader.

We made a double bias-ply covering of fiberglass-reinforced packing tape over the back side of the transfer tape and then peeled the entire pattern from the quarter panel. Sprinkle some baby powder on the sticky side of your pattern and you are ready to start shaping. The better the pattern fits your piece, the closer you are to a perfect panel.

# Chapter 6
# The Small Gas Tank Project

This project gives you the opportunity to apply many of the skills that have been covered in previous chapters: planning, measuring, cutting, bending, beading, soldering, and riveting. Plus, you'll acquire a few more new techniques to add to your repertoire. Restorers of old cars occasionally need to fabricate containers such as gas and oil tanks and vacuum tanks. The small gas tank project shows you the typical steps involved in producing a functional vessel. If you choose to make a tank, please follow these instructions to the letter to avoid becoming frustrated, and measure everything at least twice. A sealed vessel relies on the absence of gaps at all of its seams, of course, so this is no place to work by eye or intuition alone. Keep the end product; maybe you'll have an engine-testing stand some day and you'll need it for your fuel supply.

The length of our tank was chosen at random. The critical dimension is the diameter, which was chosen based on the diameter of two large sections of pipe—actually hollow steel cylinders—that we used to form the tank's endcaps. As we get further into the project, you will see how the cylinders come into play. You could easily use circular pieces of wood as long as you cut the edges very cleanly.

I made a gas tank similar to this for our engine dyno, but its construction is exactly as you might expect to find on an antique car. The skills needed to build it are a perfect culmination of material covered in this book thus far.

The size of our tank will be determined in part by the tools—two steel cylinders—that we will use to form the endcaps. Here we are scribing a circle on the tank baffle that will help locate the upper cylinder when we take the assemblage over the hydraulic press. We will cut out the semicircles marked in black before then.

We began by cutting three sheet metal circles with a diameter of 7¼ inches (radius of 3⅝ inches). We scribed an additional circle inside each of these at 6¼ inches (radius of 3⅛ inches) to mark the edge of the perimeter flange on each circle. In the first metal circle, we cut out four shallow openings around its edge; the size of the cutouts is not critical. This panel will be a baffle that will be installed inside the tank. Inside the two remaining circles we scribed another circle at 4 inches (radius of 2 inches) to mark the location for a pressed-in rib to stiffen the endcaps.

Next we sandwiched endcaps and the baffle one at a time between two large steel cylinders and centered them in a hydraulic press. With enough pressure that nothing would move, we hammered over the flange that protruded around the edge of each endcap and baffle. We hammered against a hardwood corking tool—an old oak Hupmobile spoke—to induce cold-shrinking rather than beating on the steel directly, which might have stretched the flange. On the two circles that have a 4-inch circle drawn in the middle (the endcaps), we swapped the large steel cylinder on top for a smaller one of about 4 inches in diameter and mashed an indentation into the center of both endcaps using the hydraulic press. The

indentation will stiffen the cap further and look neat. At this point, we had two endcaps and a baffle.

We cut a piece of mild sheet steel 10 inches wide by 21⅜ inches long to use for the body of the gas tank. We chose the 10-inch width arbitrarily because we didn't want the tank to be too large. The second dimension is derived from the amount of material it will take to wrap around the 6¼-inch endcaps we just fabricated. That dimension is the *circumference* of the endcap, which is the *diameter × pi (3.14)*, plus whatever metal is needed to form a seam where the two sides join. The tricky part is that the diameter of the endcap is 6¼ inches plus *two* thicknesses of sheet metal, because the outer tank skin wraps around the *outside* of the endcaps. If it were the exact same size as the endcaps, 6¼ inches, the endcaps would not fit inside the tank. To measure the actual diameter of the endcap you have to take into account that extra metal. This works out to 19.85 inches (6.32 inches × 3.14) + 0.036 inch + 0.036 inch (twice the sheet metal thickness) + 1.453 inch (metal for seams) = 21⅜ inches (21.375). The metal for seams was determined partially by what I knew we needed (two seams at ⅜ inch each) plus painful trial and error to determine the amount of metal needed for an offset between the seams.

We center the steel blanks between the cylinders and clamp them tightly together in the hydraulic press. We proceed to hammer over the flange. Our corking tool is oak because cold-shrinking usually needs one hard surface and one that is softer than the material being shrunk.

After each endcap flange is complete, we remove the top steel cylinder and replace it with one of a smaller diameter. We apply just enough pressure in the press to stamp the top cylinder about ¹⁄₁₆ inch deep into each endcap. For the indentation to come out evenly, make sure the piece is level before pressing.

Eventually we will need ½-inch-wide flanges to wrap around the endcaps. Using a sheet metal brake, we bend a flange at 45 degrees on each end of the tank body. Our flanges will be in the way until we are ready to put in the endcaps, so we flatten the flanges back out in the English wheel. We will re-bend these flanges along these same lines once the endcaps are installed.

An offset bead down the length of the tank body will give us enough clearance to lock the opposing sides of the body seam together and to hammer the seam flat.

We bent a ½-inch-wide flange at 45 degrees along both long sides of our new outer skin. The flanges were both bent in the same direction because in the end they will fold inward around the endcaps. We then ran the bends lengthwise through the English wheel using a flat wheel to remove the bends. Despite our wheeling, the memory of the metal created by the first flanges will remain so that it will bend again along the same lines later. These flanges will eventually be folded back over the endcaps.

Next we marked off and drilled four holes for screws to hold in the baffle. On the sample tank, we drilled the first hole about 4¼ inches from one end and then moved 5 inches between each of the next three holes. On your tank, feel free to space the holes perfectly so that they straddle the main seam.

To create an offset along the length of the tank so that the main body seam will sit somewhat flat, we ran an offset bead in the bead roller ⅞ inch from the edge of one of the short sides. The offset kicks to the inside of the tank so that the main seam can be rolled over on top of the offset. You will want the offset to kick inward to make room for the soldered seam. We knew which side was the inside because the flanges we bent a few moments ago were bent to the inside. It doesn't hurt to write *inside* in permanent marker on the tank, however. I have messed this up at least once.

Read or review the chapter on lead soldering if you haven't soldered in a while to get the most out of this part of the demonstration. We next applied Tinning Butter to the inside edges of the flanges that run along the longest sides of the tank, along the outside of the skin where the holes for the baffles are, along the outer edge of the metal next to the long body seam, and on the inside of the tank skin directly opposite of the body seam, because these two surfaces will eventually be soldered together. We also tinned both sides of the metal where the endcaps will go, as well as the flanges of the endcaps. We cleaned all the tinned areas with white vinegar because the Tinning Butter is corrosive.

Using a small sheet metal brake, we bent a ⅜-inch flange along the short ends of the tank skin and used a hammer to close the flanges over. The picture taken after the tank has been through the slip roll shows how one flange points

up and the other down; this configuration is so they will interlock. We used the slip roll to turn our flat blank into a cylinder. We inserted the skin between the first two stacked rollers and tightened the rear offset roller in progressive steps. I'm sure it looks more complicated in pictures than it is.

To use a slip roll, gently move the metal in and out by cranking the handle on the machine and tighten the screws on the stacked rollers until the skin rolls evenly without slipping. Tighten the thumbscrews on the offset roller so that even pressure is exerted across the width of the skin. Uneven pressure will create a funnel shape, with the tighter side creating a tighter roll. Roll the skin back and forth, gradually increasing the pressure created by the offset roller to bend the skin into a cylinder. When you're happy with the diameter of your cylinder, loosen the rollers and remove it.

Apply Tinning Butter to all of the sealing surfaces: the endcap flanges, the rivet holes, the long seam, and the ends of the tank body.

We place our flat blank in the slip roll and squeeze it evenly between the stacked rollers.

We create a cylinder by moving our metal back and forth while raising the offset roller in relation to the stacked rollers.

Here is the cylindrical tank blank next to two endcaps and a baffle. Pay particular attention to the orientation of the flanges where the tank body will need a seam.

We lock together the main seam along the length of the tank body and hammer it tight against a T-stake with a mallet.

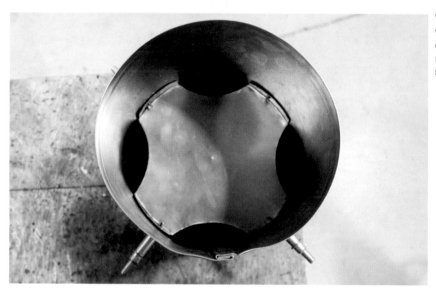

We will secure our baffle with machine screws, which we will eventually solder so that they resemble rivets, but you could rivet the baffle in place. Insert the rivets with the manufactured head on the inside of the tank and form the shop head on the outside.

To prevent the endcaps from being pushed too deeply into the tank, we need to press a half-round bead into each end of the tank body. Here is the easy way to mark off the depth around the perimeter of each tank end.

Roll the tank through the bead roller back and forth, increasing the pressure with each pass. Stop the bead on either side of the main body seam.

With our tank looking cylindrical, we pressed the flanged sides together so that they interlocked and hammered the seam flat on a large T-stake. To establish a firm seat for both endcaps, we rolled a half-round bead 1 inch from the ends the tank with the bead protruding to the inside of the vessel. If you try this yourself, do not attempt to roll across the longitudinal seam; stop on either side of it.

We inserted the baffle inside the take and lined up the holes in the baffle with the holes in the tank skin. A purist could rivet the baffle in place. We chose to attach it with machine screws that we will solder once the tank is together. The lead solder seems to hold the nuts in place on its own, but you may tack weld the nuts in place before soldering if you desire. To keep the endcaps from pushing all the way into the tank once they are installed, we rolled a bead an inch from each end of the tank.

We inserted the endcaps—they are a really tight fit— and we carefully crimped the tank body to the endcap flanges where the seam crosses. Then we folded over the edges of the tank around the endcaps. We used 20-gauge steel for our tank, but a lighter gauge would have made folding the ends much easier. We soldered the endcaps, rivets, and main seam to keep the tank from leaking. If we were planning to use our tank, we would add an outlet fitting at this point and a cap. Speedway Motors offers a nice weld-on neck with cap that would work well. Because we built this tank for purposes of demonstration, we simply drilled a hole for a fitting and leak-tested the tank by covering all of the seams with soapy water and blasting compressed air into the fitting hole. We had a leak on one end that was easily sealed with a little more solder.

Inserting the endcaps into the tank body will take some coercion. We used the same steel cylinders that we formed the flanges over and hammered the caps in place.

With the endcaps installed, crimp the body seam where it crosses each endcap to facilitate sealing.

Fold the endcap flanges over the endcaps with Vise-Grips. Proceed in gradual steps all the way around the perimeter. Moving too much metal at once will create a bad wrinkle and the tank won't seal.

Tightening up the endcap flanges cleanly requires a mixture of finesse and brute force. Hammering the caps along the inside of flange will get you far.

Placing the tank on the end of a T-stake and hammering the tank ends from the outside with a plastic mallet is another good technique for smoothing out the ends.

After the endcaps were secured, we soldered them and all of the remaining seams with 30 percent tin/70 percent lead auto body solder because it was handy. We could have used 50/50 or 40/60 acid core solder instead.

To test the tank's seal, we covered every seam with soapy water and blew compressed air into the tank through the outlet hole. Bubbles indicated a small leak on one end that was easily repaired with solder.

# Chapter 7
# Advanced Sheet Metal Shaping

As you begin to understand how sheet metal behaves, you will become anxious to get faster results. You may have seen or at least read about the staggering array of equipment now available to make your work easier and faster to produce, and you will know enough to understand how various new machines can help you. There is a mechanized tool for every metal-forming task it seems, and there is something for every budget. While a machine's price and capabilities often go hand in hand, it is possible to acquire a piece of equipment that will improve the comfort, speed, and quality of your work without spending a fortune, but don't be too hasty to start spending your money. The critical consideration is to determine exactly the kind of tasks you need to perform to create your work and shop accordingly. Before you buy anything, try to take a class or seminar from someone with equipment similar to what you are thinking of acquiring. Make sure that you can get an accurate appraisal of the tool as opposed to a blatant sales pitch. After trying out someone else's equipment, you will be much better informed when it comes time to purchase your machines. The tuition for your class will be well worth it if you save thousands on a significant tool purchase. Talk to people online in one of the metal-shaping forums. Do your research. I would not recommend getting carried away with acquisitions until you have acquired a feel for hand shaping, however. Unless you have some understanding how sheet metal moves, fancy equipment will only allow you to create scrap more quickly. I suspect that the best shapers could produce great work using only a brick, an old pipe wrench, and a sock filled with sand. The longer you extend your apprenticeship to that taskmaster Hand Work, the better prepared you will be when you move ahead into more advanced shaping.

Lance Butler decided to recreate this tattered Kurtis-Kraft quarter midget tail section in aluminum. It's still a work in progress, but it suggests that great potential shaping projects are out there if you look for them. Lance borrowed the original from a guy he met at a swap meet.

# POWER HAMMERS

Nothing inspires awe and devotion in sheet metal enthusiasts like a good old-fashioned power hammer. Unfortunately, the size, scale, and cost of a traditional Yoder-style hammer places it beyond the reach of most auto restoration folk. To learn more about the full spectrum of power hammers and power-shaping equipment, consult Timothy Paul Barton's recently published seven-volume series *Metalshaping: The Lost Sheet Metal Machines* (www.autofuturist.org). Barton's books are revolutionary in their depth and scope and are a necessity for serious students of metal working. Barton discusses the history of power hammers and their progeny and surveys all the latest iterations of hammers and metal machines, many of which are produced in scales and at prices that are within the means of small restoration shops and hobbyists.

I recently had the opportunity to attend one of Fay Butler's three-day seminars in Wheelwright, Massachusetts (www.faybutler.com). The experience was absolutely transformational, to say the least—much like compressing four years of college into three long days. For people who anticipate a career in metal shaping, attending Butler's seminar should be the first step toward becoming a professional metal shaper. Over the course of the seminar, students are introduced to the science of the materials and to the rationale for selecting certain pieces of equipment to shape and weld metal.

Butler's main pieces of equipment are two Yoder power hammers, a Pullmax sheet metal machine, and a Miller TIG welder with a special Argon purification system. According to Butler, all sheet metal shaping and welding should be dictated by proven scientific principles. Anything that does not stand up to scientific scrutiny, therefore, should be discarded. In the classroom and in the shop, Butler provides compelling evidence for the benefits of shaping through compression as opposed to hammering into soft surfaces.

Butler calls the movement that occurs when a piece of metal is compressed between two steel dies *pure* movement because all of the movement takes place above the metal's elastic limit and is therefore permanent. The elastic limit is the point at which metal cannot spring back when a load is applied. The scientific reason for this is that all of the movement above the elastic limit happens along dislocation lines within the metal's crystal structure. Hammering into sand bags or soft die surfaces, in contrast, is less efficient and less predictable because inevitably a lot of movement is simply pushing and pulling below the metal's elastic limit. Movement below the elastic limit is due to atoms temporarily moving in relation to one another. When the force is removed, the atoms return to their original positions within the crystal structure of the metal. This phenomenon of working partially below the metal's elastic limit explains why hammering into a sand or shot bag always involves hammering to get a piece close to just right, smoothing, and then additional hammering because some of the shape seems to evaporate during the smoothing process. The shape that mysteriously evaporates is shape that was created below the metal's elastic limit. Additional benefits of working above the metal's elastic limit are increased dent resistance, remarkably true surfaces, and the absence of stress concentrations that result from stretching into soft surfaces. Having experienced the benefits of shaping with power according to Butler's scientific principles, I am firmly convinced that his is the path of true professional metal shaping.

## Mini–Power Hammers

Thanks in part to the growth of interest in automotive and motorcycle restoration and customization, innovative metal folk have designed a number of smaller metal shaping machines to fill the market niche below the massive full-scale Yoder, Pullmax, and Eckold machines that have been metal-shaping industry standards for the last half-century. New, smaller machines, whose weight is measured in pounds rather than tons, can be rolled about on casters in case your workshop, heaven forbid, must occasionally serve as a garage for the family minivan. The new generation of small hammers does not rival the traditional Yoder-type hammer in ability to shape sheet metal quickly, especially heavy or resilient material, but the new hammers provide some shaping options for people for whom cost and size is more important than volume of production.

I will demonstrate the use of one such mini-hammer, known as the Dake power hammer. It is designed for up to 16-gauge steel or 11-gauge aluminum. It has an 18-inch throat so that you can reach into the middle of a 36-inch panel. It is sturdy, has a small footprint, is moveable, and uses 120-volt, single-phase power. The opening between the dies is generous enough to be able work on panels that have some curvature, which is nice. The machine comes with thumbnail shrinking dies and accepts common dies having a ¾-inch or 19mm square supporting shaft. In addition, because the shafts are square, the dies can be rotated 90 degrees for even more access when needed. The only drawback of the Dake, in my opinion, is its limited capacity for stretching, which is inherent to this type of machine with a fixed stroke. Larger power hammers, in contrast, deliver a dead blow–type hit by slinging the top steel die down against the fixed lower die. The fixed stroke machines are derivatives of the Pullmax-style sheet metal machine, which is used mostly for nibbling, shrinking with thumbnail dies, flanging, and countless beading operations. By far the most informative source of information on this type of machine is Fay Butler's book on the subject, available on his website (www.faybutler.com).

As you start shaping with power, remember that the basic principles of shaping sheet metal hold true whether the tool is powered by a machine or a limp-wristed author. A steel hammer hitting a piece of sheet metal against a steel surface will behave the same as ever, though the degree and rate of change may be drastically increased depending on the machine. Do not be overwhelmed by tool choices. As with hand shaping, there are often many solutions to a single problem. When pondering tool or die choices, think of how you would choose hand tools. For example, a die with a semisoft face, such as leather or plastic, matched with a steel die, will behave like a shot bag and a steel hammer. Although two steel dies might be the best choice in a given situation because you would prefer to stretch with compression, you may not have a machine powerful enough to achieve the results you want. If so, you may be forced to use one hard die and one soft die to make any headway. Furthermore, though there might be a best choice for each situation, do not be afraid to experiment or to try to wring a little extra finesse from the so-called slightly wrong tool. Sometimes you make great discoveries this way, and besides, you will never be able to have every tool, or even enough tools, regardless of what your family and friends may have already told you. If there is any truth to the myth of the suffering artist, it can only be due to a scarcity of tools, a shortage of work space, and the criticism of naysayers who fail to understand the boundless joy that comes with creative work.

## SHRINKING THE PIECES

As you may already know from shaping with hand tools, shrinking is the greatest obstacle to hand shaping. With enough persistence, you can make anything with hand tools, but you will be forced to plan your pieces to minimize shrinking. By dividing up complicated pieces into smaller segments that exploit stretching, you can minimize the difficulties of shrinking by hand. With a machine equipped with thumbnail dies, however, you can shrink as deeply into a panel as you wish, and shrinking becomes almost as easy as stretching. To illustrate the dramatic effect a machine with these dies can make on your work, we have installed a set of thumbnail shrinking dies and created a test piece. Similar to a hand-tucking tool, these dies create a wrinkle as the metal is inserted between them. The wrinkle is hammered flat, and therefore upset, as the metal is withdrawn. In my opinion, the usefulness of these dies just about justifies the cost of the machine even if you never use it for anything else.

The Dake mini–power hammer is modeled roughly on Pullmax-type reciprocating machines. Although much less powerful than a Pullmax, it will do a lot of work for its size and mobility.

A close-up view of the thumbnail shrinking dies. The bump on the lower die creates a lump as the metal is pushed in that gets upset, or shrunk, as the metal is withdrawn.

We pushed a panel into the machine and shut if off to illustrate what happens during the first half of the shrinking process. A completed shrink is nearby.

To set up the machine, scribble on the edge of your panel with a permanent marker and stick the panel halfway in between the dies. As you shrink more and the metal gets thicker, you may have to increase the gap a tiny bit.

As long as you keep your panel on the flat area of the thumbnail dies, you'll be able to adjust them close enough to work, but not so close that they clack together. Try to bruise the marker line without smashing the edge of the panel. The handwheel under the bottom die changes the gap between the dies.

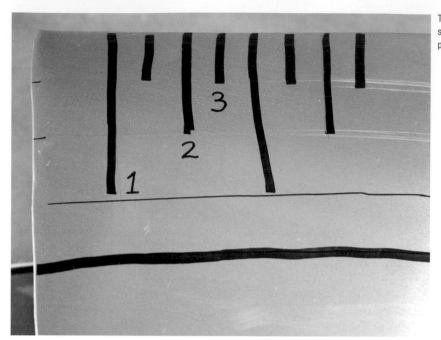

The shrinking pattern I recommend with the thumbnail shrinking dies is to stagger your shrinks: for each deep pass, make a shallower pass on either side.

Ugly for now, but well on its way to beautification, the shrunk area clearly stands apart from the virgin metal.

If you have experience in hand-forming, you will not be the least bit concerned if your first panel looks like something salvaged from the scrap bin of a seventh-grade metal shop class. At times the metal must pass through an ugly phase as it is transformed. Inevitably, thumbnail shrinking dies induce a little ugliness, but there is a huge reward for persevering nevertheless. For optimum results during shrinking, spread your shrinking over the area you intend to shrink. I believe the easiest way to accomplish this without losing your place is to make a set of passes to a certain depth from the edge of the panel. Make another set of passes less deep and on either side of the individual shrinks of the first pass. For the next set of passes, come in even less deeply than before and try to straddle the last set of shrinks. As a result of this process, the amount of shrinking increases as you approach the edge of the metal, and the metal, though lumpy, will not develop ridges clearly defining each set of shrinks.

After a few passes, areas that need additional shrinking will be obvious because the metal will form large wrinkles in those places. If you find yourself getting lost using the shrinking method I just described, try drawing two or three lines 1 to 2 inches apart along the perimeter of your piece. Make the first set of passes to the deepest line, the second set to the second line, and so forth. This technique will accomplish the same thing as the first method, but you may find that ridges will form at the terminus point for each set of passes. To get rid of the ridges, break them up with additional scattered individual passes through the thumbnail dies, or plan to smooth them out with the English wheel.

If you are shrinking to create a curved panel, I recommend gripping the metal on either side of the dies.

Modify your grip as you go to stabilize the panel. As you withdraw the panel at the end of each stroke, pull it down as you withdraw it. Forcing the metal toward the dies seems to prevent the metal from reacting to the hammering/upsetting portion of the shrink, which makes the shrink more effective. You will see large wrinkles disappear before your eyes. Pulling the panel down as you withdraw it simply directs the metal where you want it go eventually anyway. This process will create less drama in the unworked metal ahead of the area being shrunk. Do not pull down on the panel until you have started withdrawing it from between the dies, however. If you pull down on the work piece while the shrinking jaws are at their deepest point in relationship to the panel, you will create an unsightly bulge from interference of the lower die.

When shrinking around a curve, go ahead and pull down on the piece after you have started to withdraw it. The metal will be getting shorter across the dies, so you might as well get the metal headed in the right direction to follow the curve.

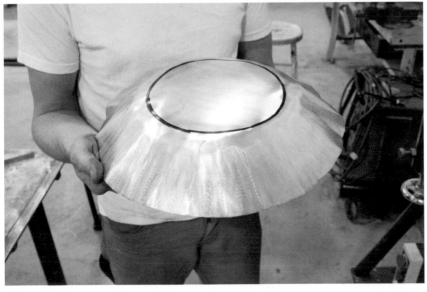

A finished round of thumbnail shrinking results in a somewhat homely wok shape.

## STRETCHING THE PIECES

Stretching is not a strong suit of a small machine with a fixed stroke such as this, but some stretching is possible. To demonstrate stretching, we made our own rubber-faced upper die and matched it with a highly crowned lower die. A highly crowned die will, of course, focus all of the available power into a small area, so the effect is similar to hitting a shot bag with the ball end of a ball-peen hammer. Just as you would when working with hand tools, however, try to work the panel evenly.

After stretching the center and shrinking the edges of the sample piece, we desperately need to smooth the surface before anyone sees it and becomes frightened or perhaps nauseated. You need not spend a lot of time in the so-called ugly phase, but we have chosen to embrace it wholeheartedly to show you that it is nothing to be afraid of. In addition, the rough stretching and shrinking we've done so far make the effects of each phase of work obvious. If you were smoothing this piece by hand, you would find a stake or dolly that matches the curvature of the finished piece and hammer the entire surface of the work against it. Remember that the distance between the dies is adjustable, so you can increase the gap for softer hits. Because of the unevenness of the sample piece, our planishing progress is easy to see and evaluate.

## BEADING

Reciprocating machines are versatile in their ability to create special edges, flanges, and beads. With the proper tooling and a good guide against which to steady the work, you can make a lot of different details. For Pullmax machines and small reciprocating machines like the Dake power hammer, consult Fay Butler's comprehensive book, *The Universal Sheet Metal Machine*, to learn how to make dies.

With a rubber-faced upper hammer and crowned steel lower die, the center of our bowl is easily raised through stretching.

Using a flat upper hammer and a crowned lower die, we planish the bowl until the surface is regular and smooth.

These beading dies are a work of art. Tapering gently at the entry and exit, they allow you to bead pieces smoothly that would never fit in a bead roller.

We adjust our fence to establish a consistent depth for our bead and sneak up on our finished bead depth by making several revolutions, reducing the gap between the dies with each pass.

The finished bead looks nice. The trick is to make several passes and hold the work steady.

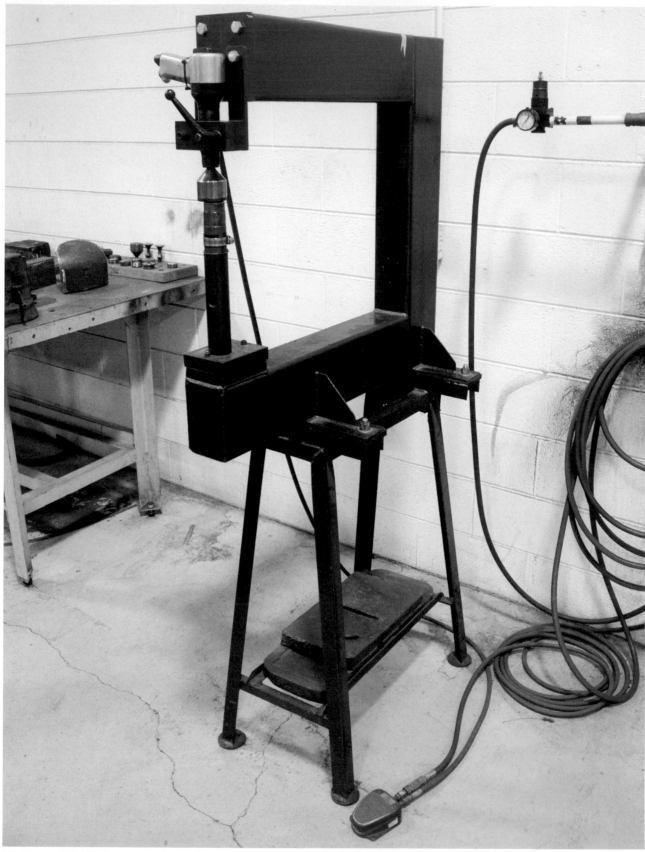

This air-powered planishing hammer comes from TM Technologies. It has seen constant use for about 10 years and has held up well.

We shrank the front of the demonstration piece with the thumbnail dies to get the leading edge of the panel to curve down.

For a demonstration of beading with the Dake hammer, we've installed a set of dies for creating a ⅞-inch half-round bead. These dies also work on Pullmax machines that accept ¾-inch/19mm square shaft tooling. To obtain the best results with these dies, we will make more than one pass through the machine, decreasing the distance between the dies each time. Holding the work piece tightly against the lower die and the fence will ensure a straight, smooth bead.

## PLANISHING HAMMERS

Modern planishing hammers are descendants of the pneumatic hand-held tools that were once used to smooth damaged automotive panels without removing them. To illustrate some of the capabilities of a typical planishing hammer, we will create a shape reminiscent of a swooping automobile fender. This particular shape incorporates a reverse curve, one of the more confounding shapes you will encounter until someone has explained to you how to achieve it. A reverse curve is composed of two opposing curves. If you work directly by attempting to bend one curve and then the other, the second curve will tend to flatten out the first curve. The key to preventing the second curve from undoing the first is to do a considerable amount of stretching while you form the second curve. Stretching the second curve allows it to develop independently from the first curve, thereby leaving the first curve undisturbed. In fact, by continuing to stretch the second curve, you can actually increase the curvature of the first curve.

We started the demonstration piece by bending the main shape of the panel by hand against a large pipe. The first bend established a relaxed, lazy S shape that runs down the center, or backbone, of the panel. Next we used the thumbnail shrinking dies in the Dake power hammer to draw the front edge of the piece down like an automobile fender.

If we had continued shrinking around the perimeter of our panel, we would have created more of a bowl shape. The shrinking would have pulled the center of the panel up, which is the opposite of the lazy S that we would like to maintain. In order not to change the backbone of our piece, we installed a flat upper and linear stretching lower die in the planishing hammer. The elongated footprint of the linear stretching die stretches the metal much more to the sides of the die than at the ends. A linear stretching die gives you more control over the direction your metal moves, therefore, than does a die that stretches in 360 degrees. The panel is inserted in the machine upside down with the linear stretching die oriented perpendicular to the backbone of the work. The piece is passed back and forth through the hammer to stretch the outer edges of panel. Remember that when stretching you are typically trading thickness for surface area. In this example, the stretching is manifested in additional length along the sides of the panel. We need this additional length to be able to curve the sides down in opposition to the backbone of the panel.

**169**

The trick to maintaining the reverse curve inherent to this piece is stretching the sides that would otherwise pull against the main curve of the panel. We use a linear stretching die in the planishing hammer with the die perpendicular to the panel for now. The clamps secure a flat bar that keeps the die from rotating in use.

Because the linear die has an oval footprint, the ovals represent the orientation of the linear die on the underside of the piece. We will be adding more length to the metal on either side of the oval footprint than at its ends.

As you begin stretching, and therefore adding length to the edge of a panel such as the one shown here, the stretched area will begin to curve up or down in response to the work. Because the metal on either side of the area you are working is fixed, you are essentially adding length between two stationary points. The extra length must go somewhere. Use this to your advantage by applying a little downward pressure against the linear stretching die. In the example here, we want the outer edge of the panel to curve down anyway, and because the panel is upside down on the linear stretching die, pushing down makes the metal go where we want it. The stretched metal will not necessarily have the contour or surface finish we want, but we can take care of those issues in subsequent steps. The first concern is to add length to allow us to bend the edges of the panel down without disturbing the backbone.

After stretching the edges of our panel with the linear stretching die perpendicular to the backbone, we flip the panel over and turn it so that the die is parallel to the backbone. Also, we swap out the flat-top hammer for a slightly curved one to avoid marring the piece as we work the curved portions. Next we work the sides of the panel, pulling down on the panel as we hammer it to create a rounded profile along the panel's edges.

With the overall shape of our panel established, we trade the linear stretching die for a crowned lower die and couple it with an upper die having a leather inset to facilitate stretching. By using low air pressure to our hammer and a light touch on the foot control, we stretch the flat area at the highest point of our sample panel. We then blend the area of highest crown into the shoulder that formerly marked the limit of our shrinking and add a little curvature to the heavily shrunk portion along the front of the panel. At this point, our piece looks much more like something made on purpose and less like an accident, but it is still a little lumpy. We planish and smooth the piece with a flat upper hammer and two differently crowned lower dies. The highest point and sides of the panel are smoothed with a die having a similar crown to those areas. The transitional areas that bridge the flat and curved sections are smoothed with a die having less crown than did the first die.

With the main stretching complete, we rotate the work so that it is parallel with the linear die. Now we will be forming the edges down. This is a combination of bending and stretching to create a downward curvature without disturbing the panel's main curve.

The topmost area of our panel is still flat, so we install an upper die with a leather insert and a crowned lower die to do a little stretching. Because it has give, the leather upper die will allow stretching just like a shot bag would if working with hand tools.

To clean up all the transition areas, we planish our panel with a steel upper die and steel lower dies of two different radii. The shape of the bottom die in each case will match the intended shape of the panel in specific places. The two areas are delineated with permanent marker.

To smooth the edges of our panel a bit more, we run the sides through the English wheel with light pressure.

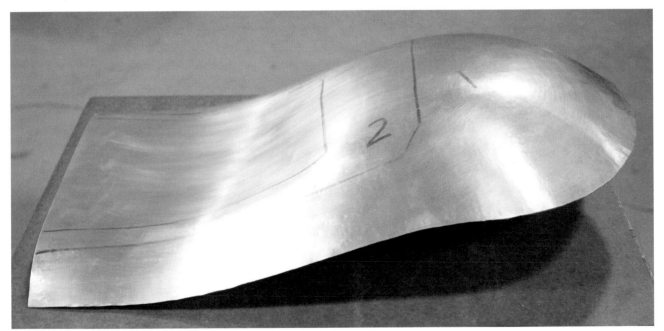

The finished panel looks good; it's reminiscent of a Ford Model A front fender.

To put the finishing touches on the sides of our panel, we run it through the English wheel with the highest crowned anvil wheel installed. If desired, you can add additional curvature at this stage by gently pushing down on the panel adjacent to the contact patch as the panel goes between the wheels. The added pressure will curve the edge of the panel. If this operation threatens the main curve of the panel, stretch the sides more with several passes with the highly crowned wheel, and then push down.

## ADVANCED DEMONSTRATION: FORD GRILLE SHELL

Now that you have been exposed to a range of metal-shaping techniques, you are ready to build a complex piece from scratch. If you are anxious to build something as complex as a fender or motorcycle gas tank, you may find it helpful to build a three-dimensional representation of your project known as a buck. If you are remaking something that is straight, but deteriorated, a buck may not be necessary.

173

Joe McCullough built this grille shell from scratch without the use of a buck. He used a reproduction shell for reference.

You can use the original to make a few templates and go from there. A buck usually requires a major investment of time; it may take almost as long to make as the piece. Therefore, to examine what is possible without making a buck, let's follow along as Joe McCullough builds a 1932 Ford grille shell.

This project came about because Joe wanted a 1932 Ford grille shell without a hole for the radiator cap in its top. Many hot rodders prefer the clean lines of this modified form of grille shell. They circumvent the problem of access to the radiator by replacing it with an aftermarket model with a filler neck that is offset behind the original cap's location. While aftermarket grille shells are available without the hole in their top, Joe wanted the satisfaction of making his own shell. Being short of funds, but long on talent and perseverance, Joe borrowed a standard reproduction grille shell from a sympathetic friend to use as a model.

Joe began his project with a few small test pieces to try out various techniques for making different parts of the shell.

I highly recommend this practice because you save metal and avoid frustration. Making test pieces allows you to solve some of the major problems you anticipate before you've even begun. Joe made a small test of a side panel, the flange that runs along the rear edge of the shell, and a portion of the luscious top of the shell.

Whenever you are confronted by an intimidating project, try to break it down into shapes that you can understand. In the case of the grille shell, the most accessible shape or collection of shapes was the inner flange that runs around the perimeter of the grille opening. A 90 degree flange is always a confidence-builder because everybody can grasp how it is made. Furthermore, anyone who has used the hand shrinker/stretcher also knows how you can induce a curve along a flange by shrinking or stretching. Within minutes, Joe created most of his first piece by bending a 90 degree flange in a strip of metal and shrinking it to follow the curve of one of the lower corners of the grille opening in the grille shell.

Try to start with something easy. Joe realized that the lower inner flange of this grille shell is just a 90 degree flange that has been curved by shrinking along its bottom.

To encourage the upper portion of his inner flange to curve over, he hit it on a stake with a linear stretching hammer. This action allowed the edge to fold over in reverse the direction of the opposite side of the panel without disturbing anything. This is the same principle at work in the planishing hammer demonstration earlier in the chapter.

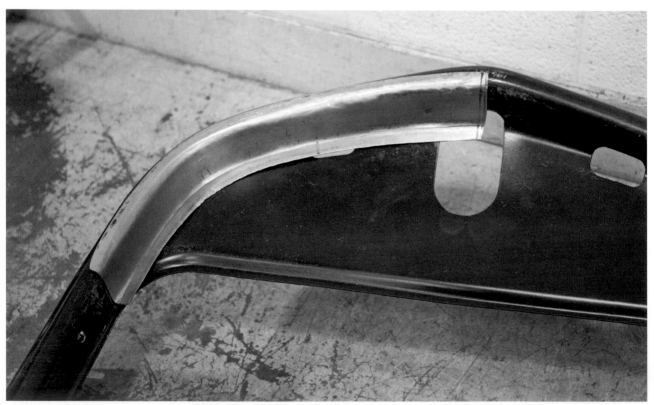

After stretching, the interior portion of the shell fits like a glove.

Metal shapers often split up a piece with a lot of shape in the area with the most shape. Dividing up the area with the most shape results in two panels needing less work than a single panel that could be insurmountably difficult to form. On the grille shell Joe divided up the 180 degree bend that runs along the front edge exactly on the crest of the curve. Shaping one half of the bend required simply hammering the top edge of the flanged piece Joe had just formed over a stake held in a vise. With a little hand stretching where the piece turns a corner, the first panel was complete. This technique was used to form the entire interior portion of the shell as illustrated in the photographs.

Bolstered by his successful completion of two or three flanged interior sections of the shell, Joe moved on to one of the large side panels, which are slightly crowned across their width and have a short bead and a 90 degree flange along their rear edges. The side panel seems long, graceful, and really difficult to make. Fortunately, this panel was very straightforward. Joe wheeled the short way across the panel to give the panel some crown, rolled a bead about ¼ inch from the rear edge, and then hammered over the short remaining tab of metal to form a 90 degree flange along the rear edge. The trusty hand shrinker was called into play to shrink along the flange and therefore curve the piece along its length. The inside edge of the panel was hammered over the same stake used earlier to make up half of the 180 degree bend that runs along the front edge of the finished shell. The hand shrinker was used again at the upper end of the panel to begin the curved transition into the very top corner of the shell.

The grille shell side panels were made by wheeling across them the short direction with a very slightly crowned anvil wheel. Joe then ran a bead near the rear edge and folded over the extra flange of metal in anticipation of making the back of the shell.

To zero in on a perfect fit in the manner of the legendary John Glover, Joe put the side panel in position and scrubbed it slightly back and forth to generate some scratches on the underside of the panel. These scratches indicate where more stretching is needed.

By now, Joe was delirious with success, which inspired exactly the confidence needed to tackle the uppermost panels, the signature panels of the shell. As described in the photograph caption, each of the two upper panels required some shrinking with the thumbnail dies along the front and hand shrinking along the edges, followed by some wheeling to smooth the panels. Each of the two upper panels could have been made by predominantly wheeling, but starting with shrinking got the shape close very quickly. Joe decided to make a separate panel with a crisp crease down its center to replace the radiator cap hole in the original shell.

Because his grille shell was composed of several small panels that would soon need to be welded together, Joe traced the grille opening of his model shell and cut out a piece of medium density fiberboard (MDF) to use as an aid in stabilizing his panels during welding. Joe drilled several self-tapping sheet metal screws into the MDF along the outline of the grille opening, trapping the flange of the small panels under the screw heads in the process. The many small panels were thus temporarily held securely to the wood so that they would not shift as they were welded. Joe welded together all of the small panels surrounding the grille opening except for the bottom panel, which was the most challenging panel to make thus far.

The large smooth panel at the base of the grille opening sweeps gently forward as it moves farther from the radiator. It also curves down at the sides. Panels like this can be a little tricky to make because one cannot easily discern where the metal is bent, stretched, or shrunk. If you find yourself

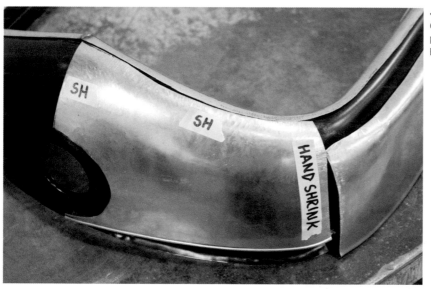

Joe began the upper panels of the shell with some quick shrinking, indicated by the *SH* taped to the panel illustrated. He then smoothed the panel in the English wheel.

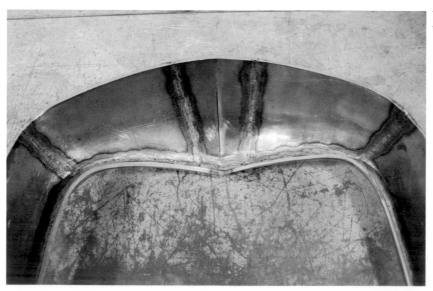

This view of the back side of the grille shell shows where the weld seams lie.

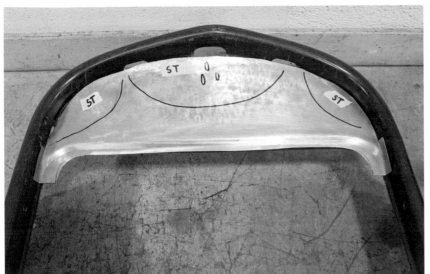

The bottom panel of the grille shell required stretching with the linear stretching die in the planishing hammer. The die was held perpendicular to the panel in each case. At the bottom center the panel advances forward, but on either side it recedes. This meant orienting the panel so that pushing against the linear die induced a curve headed in the proper direction.

Joe fabricated a low-cost fixture to secure his many small panels during welding. This technique really worked great for eliminating warpage during welding.

Joe finished the rear of the shell by stretching the edges of a U-shaped channel and running a half-round bead down the center.

flummoxed, remember that a piece of paper is always a useful guide—wrinkles mean shrinking is needed, and cuts mean stretching is needed. In this case, stretching with the linear stretching die was needed at the bottom center and at the edges to add length to the metal in those areas. Adding length forced the metal to undulate up or down according to how Joe held the piece in the planishing hammer. Once the bottom panel was joined to the other perimeter panels, Joe worked his way out, attaching the side, top, and bottom panels to his grille shell. Joe made the flange that runs along the rear of the shell by bending a U-shaped channel in the brake, curving it as needed with the hand stretcher, and finally running a half-round bead down its center.

## WIRE BUCK BUILDING

If absolute accuracy in the final piece is essential, or if you lack a solid original source piece to copy, a buck enables you to define your standard and measure your work against that standard during the entire fabrication process. A wire frame buck is the easiest buck to form quickly, but it is usually not very accurate. Made famous by the Italian coachbuilders of the 1950s and early 1960s, a wire frame buck is essentially a wire sculpture representing the outer contours of a finished car's body. In a matter of days, coachbuilders such as Scaglietti stretched out thin metal tubes over Ferrari chassis, for example, to define a car's shape for the benefit of panel beaters, who were charged with making individual panels to

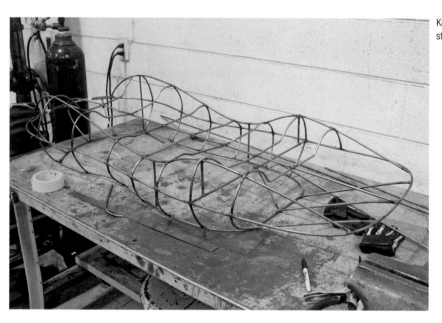

Kalila Haddad built the wire frame buck for our miniature streamliner out of ¼-inch-diameter mild steel rod.

You can see in this view of the lower front sheet metal that the wire buck provides great visibility to check the fit of your panels as you go.

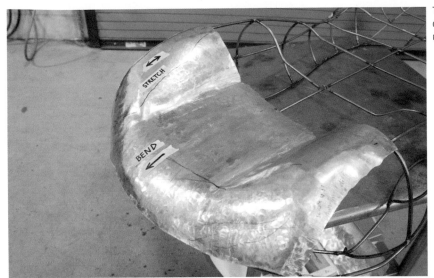

The front sheet metal consists of a bend along the front of the car with lots of stretching at the sides to allow the metal to flare up over the front wheelhousings.

Another benefit of the wire buck is the ability to clamp and weld panels in place whatever final welding process you use.

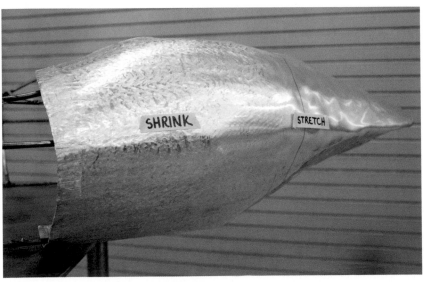

The rear quarter areas were roughed in with shrinking at the front, denoted by *SHRINK*, and stretching toward the rear, indicated with *STRETCH*.

The streamliner's top panel was almost flat, but some stretching was needed to create transitions for the wheelhouses.

We cut out a hole in the streamliner's top panel for a cockpit and created a recess for the cockpit cover with a pneumatic flanging tool.

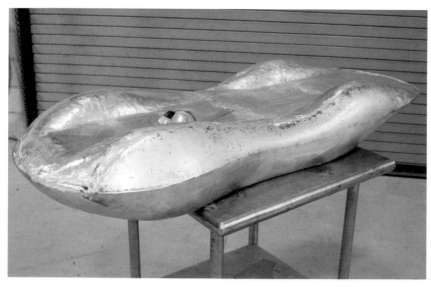

We didn't get to finish out the panels as much as we had hoped, but we welded them together, ground down the welds, and ended up with something that looks authentic to the 1950s.

finish the body. As a representation of this technique, Kalila Haddad built a wire frame buck of a high-speed streamliner in a reduced scale. Glenn Herman and Stuart Mitchell built the aluminum body for the car. The chief benefit of the wire frame technique is its speed. A wire frame buck is also useful during the welding phase because pieces can be easily clamped and welded while attached to the buck. Using ¼-inch-diameter steel rod, the contours of the entire length of our car were determined, clamped with vise grips, and then changed as needed. To speed up construction considerably, we should have run the rods that cross the body horizontally on top of the longitudinal rods that run the length of the car, rather than cutting out small sections in between. The finished buck was not exactly symmetrical from side to side, but neither were the old Ferraris, so we forged on and were pleased with the results.

## STRYROFOAM AND WOOD BUCKS

If great accuracy is your goal, consider making a wood buck based on a Styrofoam model of your project. For a demonstration piece, I made a drawing of a make-believe pontoon-style fender on graph paper. We scaled up the drawing, transferred it to a sheet of medium density fiberboard, and cut out the silhouette of the fender with a jigsaw to serve as the spine of the model. We then glued several layers of builders' Styrofoam insulation to the spine using 3M Super 77 adhesive. Using an electric grinder, a hacksaw blade, and a surform, we began shaping the foam to the finished size. The grinder removes material at an alarming rate. It reminds me of a snowblower removing snow, so be careful. Also, grinding Styrofoam is surely one of the messiest jobs in the world. Wear safety glasses and a dust mask, have a working vacuum cleaner nearby, and make sure your spouse

To build a foam buck, glue a few layers of scoreboard insulation with 3M Super 77 spray adhesive in the rough shape of your intended project.

Carve your foam to shape with a grinder, a hacksaw, or a surform.

Once the outer contour of your buck is correct, tape off and remove individual slices that can be translated into wood stations.

We used the foam buck sections to determine a rough angle for the jigsaw to minimize trimming and sanding on the wood stations.

Although we didn't use it on this project, we have a hot wire made from a battery charger that enables us to slice foam bucks of unlimited sizes.

is out of the country if you must carry out this madness at home. Fortunately, it isn't the end of the world if you err and remove too much material. Just cut out a section leaving clean sides, and glue in a new hunk of Styrofoam. Shape it as needed. You might use a contour gauge to match up your repair to an opposing side if you need symmetry.

After the Styrofoam was shaped to our satisfaction, we cut it into slices with a hacksaw blade and meticulously copied and cut matching pieces of MDF to use as stations on our buck. Although we didn't use it on this project, we have used a homemade hot wire that we constructed out of 0.035-inch-diameter stainless-steel MIG wire stretched inside a wooden frame. With a battery charger adjusted to 6-volt manual and attached at opposite ends of the wire, this tool cuts effortlessly through Styrofoam.

Before attaching each station to the buck, we cut out a large portion of the center of each station. This will make the buck lighter, but more importantly, it will give us clamping access during welding and allow us to check our fit against the buck visually as we go along. We attached the individual stations to the wooden center spine of the buck with wood glue and screws. Predrill the MDF or it will split when you drive screws into it. Although MDF is not particularly rugged, it sands beautifully. To finesse our surfaces into perfect harmony, we took a long sanding block with 80-grit paper and quickly whittled down all the surfaces so that they were accurate and consistent with one another. Lay a flat strip of steel or a welding rod across several stations to check their relationship to one another periodically.

Plan ahead and you might be able to combine the shaping purpose of a hammerform with the shape-checking purpose of a buck. By adding extra stations in areas where we anticipate a lot of shrinking, we'll be able to hammer the metal down against the buck where we've reinforced it.

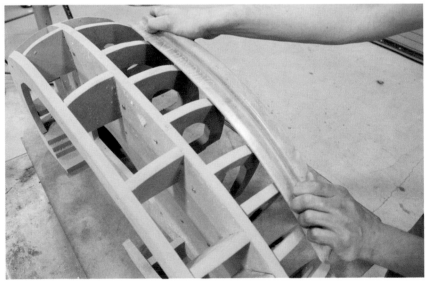

MDF cuts and sands easily. After our MDF buck is built, we even out all of the contours with a large semiflexible sanding block.

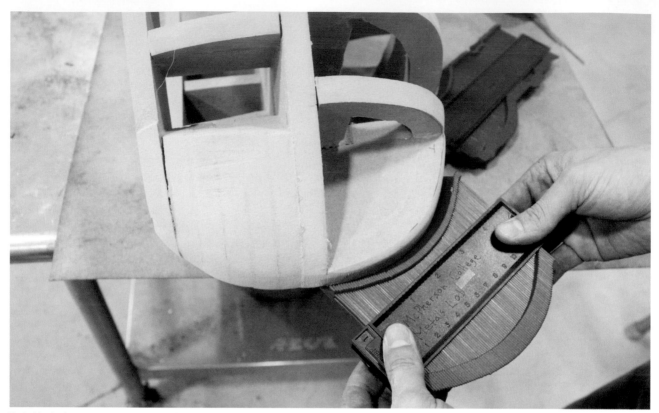

To double-check our symmetry, we use a profile gauge on the tail section.

By clamping a flat strip of metal or a welding rod across several stations, we can see at a glance if any stations are high or low in relation to their neighbors.

# Chapter 8
# Building a Fender From Concept to Completion

To demonstrate the use of the Dake hammer, planishing hammer, and other tools on an elaborate project, we built an aluminum fender based on the buck we completed in the last chapter. This description follows the construction in some detail, but that is the only way to follow the steps involved and have them make sense. Solving problems as they arise is a necessary part of learning to shape metal. Hopefully, watching us cope with different situations will offer ideas that you can apply to future projects. To get an idea of what would be involved in shaping the metal for this piece, we pinned some paper onto the buck with thumbtacks and folded and cut the paper as needed to get the paper to fit. The paper will serve as a very accurate plan for the project.

## EARLY CONCEPT AND CONSTRUCTION

One feature I was determined to include in this fender was a crisp body line that travels down the center of piece, but fades into the sheet metal at its ends. As I pondered how best to approach the body line problem, I considered joining the two sides of the fender at the crease, but thought better of that option because I doubted that we would be able to finish out the weld bead satisfactorily. We could also have made one fender half a little longer and placed the bead near the edge of that half. This procedure would have required shrinking the short side flange that would have protruded next to the crisp body line, because the body line would be the highest point of the fender. This second option would

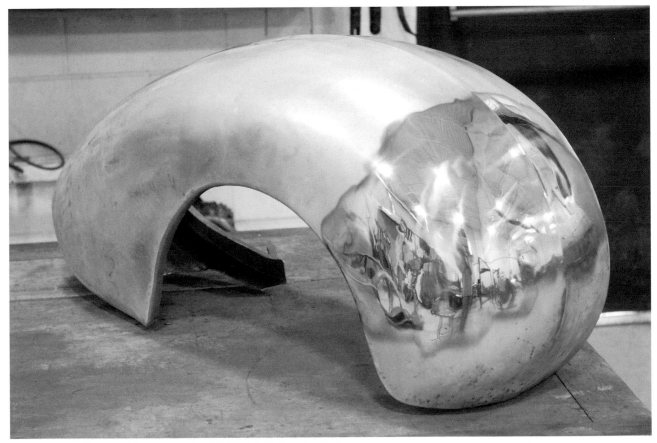

In this chapter we will go through the necessary steps to build this fender from start to finish.

Paper makes a reliable road map for a metal project. The mood lighting here is intended to highlight the contour changes of our fender.

have reduced the amount of welding required by a lot, but I was not sure we would be able to get the short flange to sit so that it would match the fender on the other side of the body line. A third option would have been to make a set of Delrin dies for the Dake and put the body line in after the fender was complete. Running a straight bead that fades at its ends on a large curved piece sounded daunting, especially if I happened to be working alone at the time. A fourth option would have been to modify the top wheel on an English wheel, either by wrapping an inner tube around it or by swapping it for a huge hard rubber caster and running the finished piece through using a sharp creasing anvil wheel. Coming up with the sharp lower anvil wheel sounded like a hassle. A fifth option would have been to build the complete fender and chase the crease into the top of the fender from the back side with a blunt chisel. I wasn't sure we would be able do that cleanly, so instead I chose to roll a bead in a separate strip of aluminum, attach that strip to the buck with Clecos, and proceed from there. This method worked really well for keeping everything aligned, but it required a lot of welding and finish work. Having the center strip firmly attached to the buck was helpful because it provided a central spine that was fixed.

Having a fixed point is useful because something is decided. If you have trouble making a commitment regarding the permanent shape of a piece, your progress will be slow. You may find yourself juggling several nice pieces that fit the buck individually, but not together, because you will be constantly shifting something in pursuit of the best possible fit.

After rolling the bead in the center strip with the bead roller, we shrunk the edges on the hand shrinker to convince the strip to curl in the shape of the buck. Once we attached the strip to the buck we began working on the first panel. We used 3003 H-14 aluminum 0.063 inch thick for this demonstration. Ironically, the first panel was made from an unmarked scrap that was clearly something other than 3003 H-14. It may have been 5052. In any case, it was quite springy and recalcitrant, so we abandoned it after the first round of shrinking. Let this be a lesson: always clearly mark your metal as it is cut from larger sheets. For the remainder of the project, we used 16-by-24-inch sheets left over from another project. Had we been willing to cut fresh metal, we would have used fewer pieces and probably would have relocated a couple of seams, but I thought using what was available was more true to life.

By putting our crease on a separate strip of metal, I have chosen the safe but laborious way to get the overall effect I wanted.

With the correct metal in hand, we laid the paper pattern over the new metal blank and used a pushpin to transfer the line indicating the limit for shrinking onto the panel. By poking through the paper, we created a line of prick marks that we then traced with a permanent marker on the metal. Because the Dake thumbnail shrinking dies work so well, we anticipated forming most of the panels on this project through shrinking. If we were making this first panel with hand tools, in contrast, we would have added a seam along the apex of the long curve. A seam in that location would require only a small amount of shrinking on the edge of the large almost flat portion of the panel, whereas the curved section of the panel could be made predominantly by stretching with a little hand shrinking on the edges.

Depending on your technique, thumbnail shrinking dies can have a tendency to leave a lump at the end of their travel into a panel. This lump becomes more pronounced the more abruptly the panel curves down. If you are fortunate enough to have a deep-throated shrinker with gathering-type jaws, eliminating these lumps is a simple matter of additional shrinking. Gathering jaws are the other common type of shrinking tool other than thumbnail dies. Like the jaws in hand shrinkers, gathering jaws grip the metal and upset it as the jaws move toward one another. Fortunately, with aluminum, lumps left from thumbnail shrinking can be cold shrunk with a wood or rawhide mallet, they can be shrunk with soft dies in a planishing hammer, or they can be smoothed with the English wheel if they are not too pronounced. Metal shaper

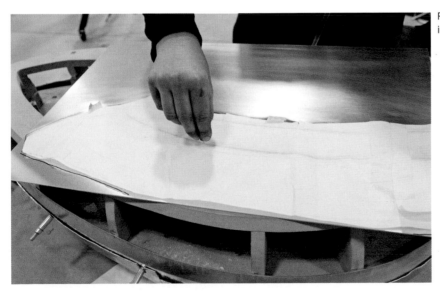

Poke through the paper pattern with a pushpin to transfer important details like shrinking limits to new metal.

Fay Butler avoids creating lumps with thumbnail dies by manipulating the form of the panel as needed so that there is no abrupt transition between the end of each shrink and the rest of the panel during shrinking. After each round of shrinking is completed, he manipulates the form back into the panel and checks it against the buck. With the panel in question, therefore, this would have meant curling the panel up like a potato chip perpendicular to the highlight line, shrinking, and then pulling and pushing the panel back into its original configuration for checking against the buck.

## FINISHING THE FIRST FENDER SECTION

After shrinking and smoothing, the first fender section looks good, but the extreme front corner needs still more

shrinking. Fortunately, we fortified the front and rear of the buck ahead of time with additional stations in anticipation of the trouble we would have shrinking in those areas. With so many stations in this area, the buck will double as a hammerform to aid shrinking. For spot shrinking, you will want to try spot annealing. Remove your piece from its buck if necessary, scribble over the area you intend to shrink with your permanent marker and then burn off the marker with an oxyacetylene torch adjusted just as you would use for welding. Let the area cool, and it will be annealed. Annealing such as this allows the aluminum to shrink like a dream with a soft hammer or slapper. During annealing, the effects of work-hardening are removed as new strain-free grains form in the annealed area.

Permanent marker lines indicate the general plan for shrinking we will use with the thumbnail dies.

After wheeling the lumps, the top portion of the panel fits nice and close to the creased panel.

The front corner of the first panel protrudes out from the buck and needs additional shrinking. We will scribble on the bulged area with a permanent marker and burn off the ink to anneal this specific area.

We play a welding flame over the inked area on the front panel until the ink disappears to soften the work-hardened metal in preparation for cold-shrinking.

We use a leather-covered wood slapper to spank the bulge on the front panel into submission and the fit is much improved.

Like the first panel, the second panel is shrunk along its outer edge with the thumbnail dies. Because the transition from the flat portion to the curved portion of the panel is abrupt, the lumps at the upper limit of our shrinking are quite pronounced. We place a steel dolly on top of a stack of shot bags and shrink the lumps with a rawhide mallet. This would not be so easy were we working with steel, but it is possible to heat shrink lumps like these and wheel them smooth afterward. A little more shrinking is needed along the rear edge of this panel and at one spot along the wheel opening, so we make one additional pass through the thumbnail dies. Our shaping is going well, but before we get carried away we make some registration marks on the back side of our panel to establish its semipermanent relationship to the buck.

This close-up view of the second fender section reveals lumps left from the thumbnail dies.

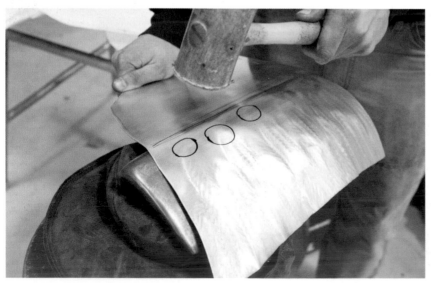

Aluminum is the best material for learning metal shaping because it is so ductile. Cold-shrinking high spots with a rawhide mallet is easily accomplished by hammering the lumps over a dolly on a stack of shot bags.

As our second panel gets really close to fitting, we make some registration marks between it and the center piece to locate its ideal placement.

The third panel, which makes up the tail of our fender, is shrunk around its perimeter with the thumbnail dies. Because of the amount of shrinking involved, this panel is the most labor intensive of the project. In situations requiring several rounds of shrinking, do not to hesitate to anneal your metal frequently to keep it from becoming work-hardened and possibly cracking. A less laborious solution would have been to put a seam at the apex of the most drastic shape, but we were curious to see if we could successfully shrink this much so we chose the difficult route. Following several bouts of shrinking and annealing, the rear section almost clings to the buck. Using the same technique we used at the opposite end of the piece, we rap the metal a few times with a rawhide mallet to tighten up the fit. The rawhide mallet works similarly to the slapper, though its face is smaller, and it was selected simply to reiterate that there are often many good solutions to any problem.

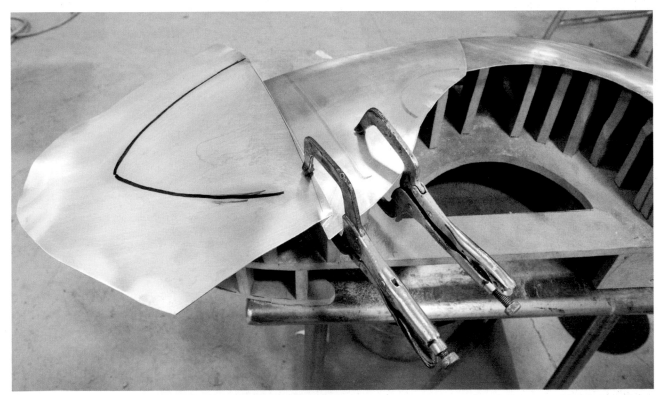

The third panel will need to be shrunk an incredible amount to lie flat against the buck.

After several rounds of shrinking, alternated with annealing, the third panel almost fits.

The rawhide mallet is called into service to cold-shrink the third panel against the rear of buck. Good thing we added so many stations to the buck in this area.

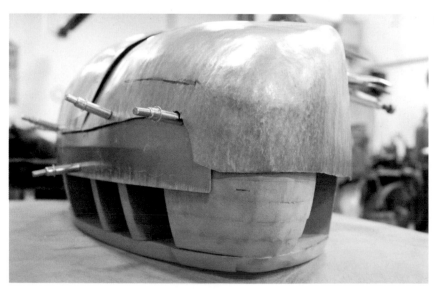

At last the third panel fits snugly against the buck.

The back side of the sample fender juxtaposes a long curved section with a wide flat flange that must eventually be bent at 90 degrees for stiffness. We shrink the first blank using the thumbnail dies without annealing the metal beforehand. Unfortunately, the panel curves upward from the tension created by shrinking. By annealing the metal and enlisting the help of a muscular student, we easily flatten the panel against a workbench. I designed the buck for this project with the goal of using it as a hammerform in some places, such as along the back side. By sandwiching the first rear panel against the buck with several C-clamps and pieces of wood, we are able to shape the first panel by hand with a rawhide mallet. Our hammering leaves the piece a bit rough, so we remove our clamps and take the piece to the planishing hammer. We use a top die with a plastic face and two different lower dies for smoothing. First we use a gently crowned lower die on the curved portion of the piece and a large flat steel die for the flange.

After annealing the recalcitrant rear panel, we enlist the help of a strapping college student to flatten the bowed panel against a table.

After smashing the panel on the table, its fit is much improved against the buck.

More shrinking is needed along the bottom of the panel, so we clamp it to the buck with several pieces of wood and hammer like we have done with previous panels.

To flatten out this troublesome panel, we sandwich it in the planishing hammer between a plasic-lined top die and a flat bottom for the flat side of the panel and a slightly curved bottom die for the crowned side.

The path you follow when planishing doesn't matter as much as that you planish evenly.

After planishing, the first rear piece fits the buck well and has a nice surface finish.

The second rear panel is created much like the first, but my stinginess with regard to materials has come back to haunt me. I remember when we traced the paper pattern onto the blank for this panel that there was no way the piece could be squeezed out of a 16-by-24-inch piece of aluminum. The dearth of material is now plainly evident where the panel fails to meet the body line that runs down the center of the buck. I had hoped that somehow this would not come to pass, but of course it has. If you ever find yourself in this situation, cut a piece of material larger than you need and grind the edge to match the missing section perfectly along

the seam to be welded. Clamping a piece of copper behind each end point for your patch will allow you to weld right up to the edge without burning a large hole. Leaving the patch oversized will help distribute the welding heat and will greatly facilitate welding. Your natural tendency may be to fabricate a patch the size of the piece you need for any given repair. When working along an edge, however, always leave extra metal past the edge. Only after you have welded the patch and ground it smooth should you trim it to size. Leaving the repair oversized will help prevent you from overgrinding the edge and will make trimming easier too.

We complete the second rear panel much like the first, but it is undersized, as indicated by the gap between it and the creased panel running down the backbone of the fender.

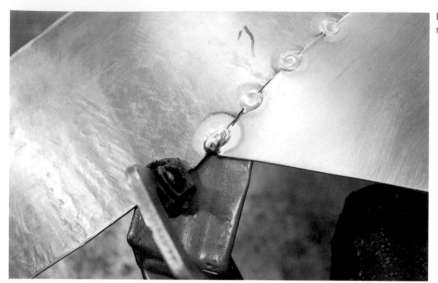

Backing our joint with a copper spoon allows us to weld right up the edge of our patch.

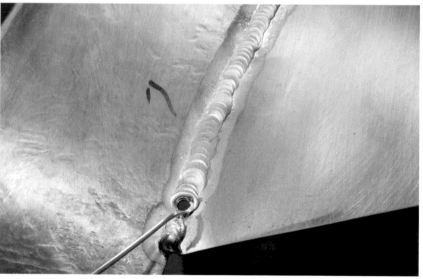

Any time you must add a patch to the edge of another piece of metal, always leave the new piece long until after you've welded it in place. Grind the weld, and then trim the patch. When welding, try *not* to let the puddle cool prematurely and make a crater, as I did.

The repaired edge is indistinguishable from the surrounding metal.

The rear edge of this panel has strayed from the buck about half an inch. To rectify this situation, we need to shrink at the area marked *SH* and stretch below it in the area marked *ST*. The shape is analogous to a lobster claw about to be drawn together.

After shrinking with thumbnail dies and stretching with the linear die in the planishing hammer, the panel fits great.

The second rear panel looks nice but does not quite fit the buck. Either one end or the other fits perfectly, but not at the same time. To remedy this situation, we shrink the flat flange and stretch the curved area adjacent to it with the linear stretching die in the planishing hammer. The combination of shrinking and stretching pulls the panel tight against the buck.

The final rear panel on the underside of the tail section is shrunk differently than the previous panels were. We decide to run the panel through the thumbnail shrinking dies parallel to the bend line where the shrunk part of the panel meets the flat flange. Normally we would shrink with the dies perpendicular to the panel. The new technique works remarkably well for shrinking right up to an abrupt change of direction without leaving any lumps near the bend line. We finish shrinking below the first pass in the conventional manner with the thumbnail dies perpendicular to the panel. We clamp the piece to the buck, hammer it down with the rawhide mallet, then planish it smooth as we have done on all the previous panels.

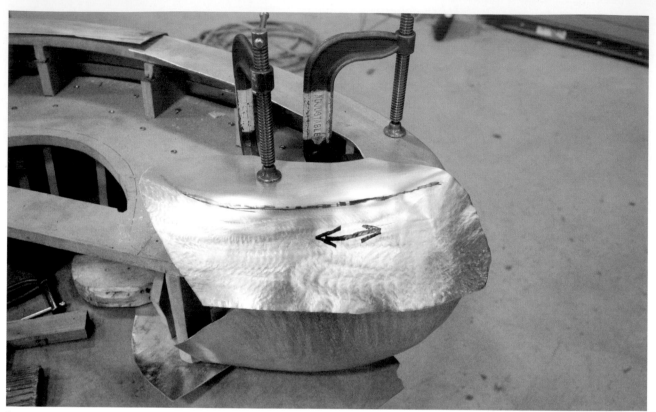

To encourage more drastic shrinking on this panel, we run it through the thumbnail dies parallel to the bend line.

As we have done on other panels, we clamp this piece against the buck and hammer it over.

The last rear panel finally fits the buck and has a smooth finish.

## WELDING THE FENDER

Having completed the last rear panel, we are almost ready to weld our fender together. Before we start welding we need to check the fit of each panel and trim them as needed. Finalizing the fit before welding is critical to avoiding large gaps, warpage, loads of grinding, and other unforeseen mishaps. Do not rush this step. The rigid center piece that carries the body line and the elaborate buck are both great sources of stability during the welding phase of this project. In addition, we intend to leave the piece on the buck until the last possible moment because it greatly reduces warpage during welding.

This is the kind of fit you should shoot for when preparing panels for welding. A precise fit is 90 percent of a weld, especially on aluminum.

201

Look closely at the very fine burr on the edges of these panels left by the grinder. File these burrs off prior to welding or they will burn up and form a weld-thwarting oxide that will mystify you. If you don't remove them, the edges of your panels will heat up, but because of the debris left by the burr, the molten edges will not run together.

Whenever possible, clamp your work to the buck to prevent warping during welding.

With our fender welded with the exception of one rear panel, we must make some last-minute adjustments before we remove the project from the buck. The final unwelded section does not quite fit as well as we would like. The unwelded panel is a little low compared with the panel that it is supposed to meet. The low spot is circled with permanent marker in the photograph. By working this area with the linear stretching die in the planishing hammer, we raise this spot perfectly. Turn the die perpendicular to the panel's edge in situations like this to stretch the edge of a panel while keeping the contour of the stretched area consistent with the rest of the panel.

Another issue we need to resolve on this last panel is its finish. Once the panel is welded in place on the fender, it will be much more difficult to planish properly. Therefore, we smooth the panel with the planishing hammer using dies having three different radii. We use the most curved die on the most curved section of the panel and change dies as we work our way back to the almost flat portion of the piece.

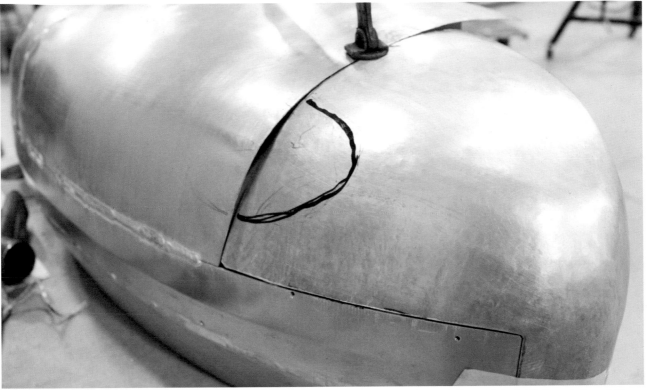

This photo illustrates a classic problem that can be solved with a linear stretching die in the planishing hammer. The poor fitting panel is too tight in the marked area. Adding length to the area will improve the fit.

We stretched the problem area with the linear die perpendicular to the edge of the panel, thereby adding length to that spot. With a little more length along the edge, the once low panel fits.

Once the final panel is smooth, we remove the fender from the buck and weld the remaining panel in place. This sounds much easier than it was, however, because we had some difficulty clamping the loose panel in its proper place in relation to the jiggling, unsupported fender. Had we made the underside of the tail section a single panel, we might have been able to keep the fender on the buck a little longer to ease welding. Once the fender is welded together, we run the TIG torch along the back side of all the weld beads to ensure complete fusion and to flatten the beads prior to grinding.

We alternate between a 4½-inch electric grinder and a vixen file for the first stage of weld finishing. When grinding or filing welds, be mindful of your terrain. As long as only the weld bead is shiny, you are in no danger of thinning

adjoining areas. As soon as your tool makes contact with the metal alongside the weld bead, however, you must pay close attention to your progress to prevent thinning the metal near the bead. Finishing a weld bead so that it is indistinguishable from the surrounding metal will inevitably involve some hammering and leveling with a hammer and dolly or perhaps a planishing hammer to minimize unevenness left from welding. Finishing a weld bead completely requires grinding the back side of the bead as well so that it does not interfere with your efforts to smooth the panel. A weld grinding wheel in a pneumatic cut-off tool works well in confined areas. As soon as the welds are ground or filed level with the rest of the fender, we smooth them as much as possible in the planishing hammer. Beads that are inaccessible with the planishing hammer must be finished by hand.

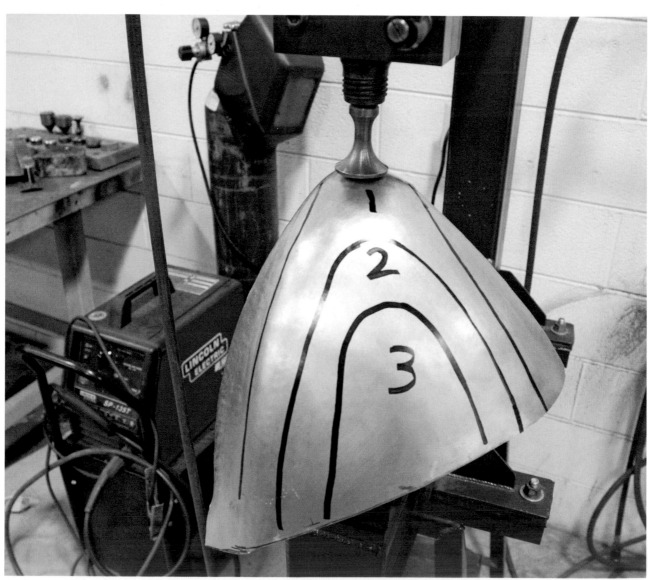

The third rear panel has a lot of shape. Smoothing this panel requires using planishing hammer dies of three different radii.

This is what the back side of our welds looks like. We like to run the TIG torch down seams like this to help blend in the bead with the surrounding metal.

An electric grinder works well for taking down the proudest weld beads. A coarse vixen file works as well.

A pneumatic tool with a weld-grinding wheel is handy for hard-to-reach places such as the inside of the fender.

After grinding down the tallest welds until they are almost flush with the surface, we planish the welds in the planishing hammer to flatten them.

Once our weld beads are ground down, we trim away the extra metal around the wheel opening of our fender and along the back side. At this point I realize that I have goofed by failing to bend over the back-side flange before removing the piece from the buck. With the fender on the buck, bending the 90 degree flange that follows the length of the piece would have been a simple affair. I designed the buck with the intention that the top would be used as a hammerform to make the flange in question. Instead, we bend the flange along the back side with Vise-Grip pliers and clean it up with a hammer and dolly. We also bend the flange along the wheel opening with the flange turning dies in the planishing hammer as far as possible and then finish the work with a hammer and dolly.

We trim off the extra metal around the wheel opening with a pair of aviation snips.

We bend the flange on the back side of the fender by placing a dolly next to the flange and hammering over the unsupported metal at 90 degrees.

We bend most of the flange on the wheel arch opening using the planishing hammer with a flat bottom die and a plastic-lined upper hammer. We clean it up by hand.

With our flanges bent, we resume the finishing process by sanding the fender with 180-grit paper on a soft pad sander. Any high and low spots discovered at this phase must be rectified with a hammer and dolly, a bull's-eye pick, or whatever means are necessary to smooth the metal enough that it can be sanded without risking thinning the metal. To further refine the surface, we go over the fender with 320-grit paper followed by 600-grit paper mounted on a dual action palm sander. The gentle random orbit of this tool eliminates gouges and deep swirl marks that can result from a tool with a fixed rotation. For the final touch, we polish a portion of the front of the fender with Tripoli abrasive on a ventilated buffing wheel. Additional sanding will be needed if we hope to achieve a flawless finish over the entire piece, but even the small amount of buffing we've done so far reveals the high level of finish possible with enough elbow grease.

To refine our surface finish, we lightly spray the tail section with a guide coat of self-etching primer and briefly block sand it to reveal highs and lows.

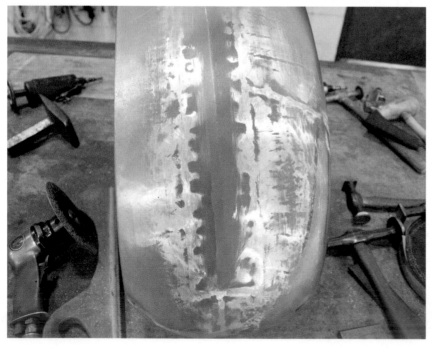

By bumping down the highs and raising the lows, we get a little bit closer to an even, flat surface that will ready for polishing.

Using the slapping spoon to strike directly on the dolly in the low area within the ink border, we will raise the metal to level, but hopefully no further.

A couple of raps with the slapper, some scuffing with the sanding block, and the primer is gone. The previously low spot is now even with the rest of the panel.

Here is a perfect application for the bull's-eye pick, a small low spot.

After a couple of careful taps with the bull's-eye pick and a quick pass with the sanding block, the low spot is much improved.

The soft pad sander is just like a sanding block that happens to be rotating. It will remove metal and help to flatten areas that are very, very close to flat. You can't remove too much material, obviously, or the metal will be thin, but this step is necessary to make the surface uniform.

We finish sanding with the DA palm sander and then buff a portion of the front of our fender to demonstrate what level of finish is possible.

# Chapter 9
# Making It Beautiful: Straightening, Grinding, and Surface Finishing

The process of beautifying a sheet metal object can be an arduous but rewarding task. If an object is corroded, dented, or incomplete, you have the opportunity to reverse the ravages of time and restore it to its former glory, but such miracle working seldom comes easily. In this chapter I will describe some techniques for removing old paint and oxidation, straightening sheet metal, and improving its surface finish to the level you desire.

## REMOVING PAINT AND OXIDATION

Depending on the goal for your project, the presence of old paint and rust or corrosion may significantly affect the feasibility of your plans. Determine the true condition of what you have as early as possible. I have seen more than one vehicle purchased in an online auction that turned out to be a complete disaster beneath layers of carefully sculpted plastic filler. Unfortunately, after the owners had spent years amassing parts in preparation for the restoration of their dream car, the painful truth became known when the owners had their cars media-blasted or chemically stripped. Find out the condition of what you have and then find out if it is worthwhile to you to move forward. The time-tested means for removing old paint and oxidation are through sanding, the application of chemicals, and abrasive blasting.

It is hard to believe that this pedal car header was made out of rusty bicycle handlebars and a length of automotive exhaust pipe. The transformation involved some welding, sanding, polishing, and nickel plating.

This Jaguar XK120 looked OK from about 10 feet. The owner purchased it for an obscene amount of money in an online auction. After the car was chemically dipped, the painful truth of its condition was revealed. Equally disturbing, in my opinion, is how areas such as this somehow escaped the chemical bath.

The easiest to commence, but perhaps messiest method of removing old paint and/or oxidation, is through sanding. The benefits to sanding are that a lot of fancy equipment is not needed, and you can deal with one piece at a time to make the project more manageable. Sand a piece, clean it, epoxy prime it, set it aside, and move on when it is convenient. For the do-it-yourselfer or person with small objects to clean, sanding is convenient. Large 7- or 9-inch body grinders with coarse 36-grit paper will give you the quickest results, but smaller sanders will certainly do the job if that is all you have. The drawbacks are the clouds of potentially hazardous dust that sanding generates, the potential for damage through warping and overgrinding if performed improperly, and the difficulty of doing a thorough job in tight spaces. Sanding is also a lot of work compared with chemical dipping and blasting.

Chemical dip stripping is performed by specialists in certain parts of the country. It may not be available in your area, so you must factor in the cost of transporting your project

to and from the facility. Dip stripping removes absolutely everything, so be prepared to apply body solder to joints that require it if you are dealing with an old automobile. If you are considering dipping as an alternative, get references for the company you anticipate working with to find out how thorough their stripping process and the cleanup procedure that must follow are. If you are stripping a large part without a lot of nooks and crevices to trap caustic solutions, this process might be a good choice for you. If there is a danger that the caustic solution used to remove the paint might not be completely removed, you could have trouble with additional corrosion and paint lifting later. I have worked on cars that were dipped as soon as they arrived at our shop. Inevitably, there were cavities that were not clean, I presume because air became trapped and prevented the solution from reaching them, but the process was extremely thorough on all exterior surfaces. Once your project is dipped and thoroughly cleaned, be prepared to have it primed immediately to avoid corrosion getting a head start on the fresh metal.

Chemical paint strippers are available that allow you the convenience of dealing with items on a piecemeal basis, but these take time to apply, they are a little messy, and they are more work than abrasive blasting. I remember using a product called Mar-Hyde Tal-Strip over 25 years ago on my first automotive project. We used the updated Mar-Hyde Tal-Strip II on aluminum panels in recent years in the restoration shop where I worked restoring British cars. This product is probably available at your local auto parts store, and it is easier and less messy than sanding. Brush or spray the Tal-Strip II on the surface, allow it to sit a few minutes, and then scrape it off. I have never used the aerosol version of Tal-Strip II, but the gel version easily removes old paint and body filler, though several coats will be needed to take care of thick layers of material. If working with softer metals like aluminum, use a plastic scraper so that you won't damage your panels. Once you are finished stripping a single panel, thoroughly wash off all the residue. If you are working with steel, I recommend

following a metal conditioning procedure such as the one I describe later in this chapter and prime the metal as soon as possible afterward. Metal conditioners are available for other metals as well, but because steel seems to rust before your eyes, I would be extra vigilant with it.

My favorite method for stripping and cleaning metal is abrasive blasting. It is fast, remarkably thorough, and a range of potential blasting media is available to clean just about any surface safely. In my experience, cars that have been blasted and epoxy primed right away are a joy to work on. The rust is gone, and each panel is pickled for eternity in a protective coating that can easily be sanded off for purposes of making repairs to panels a piece at a time. Furthermore, by blasting the project first, all hidden secrets are revealed. Blasting cabinets are available in many sizes, so this might be the ideal option for you if the scale of your work will allow it and you have or can get an air compressor with sufficient capacity for blasting. For larger pieces, seek out a blasting professional.

This Ford wheel shows how thorough abrasive blasting can be, especially on intricate parts that would a nightmare to clean by other means.

If you ever acquire a blasting cabinet, you'll wonder how you ever lived without one. We added a large sheet metal extension onto our cabinet so that we could fit large objects like fenders inside. The extension is just 20-gauge sheet metal riveted to the cabinet with some sealant at the seam. It slopes down at the bottom to facilitate abrasive return.

Because abrasive blasting is popular and accessible, you will likely find several professionals near you. Do your homework before you entrust anything to anyone, however. Consider having your chosen shop work on something small before handing over your prized project. You might even send them a door from a junkyard. Ask for references and look at examples of the shop's work. I vividly remember a flat panel that came back from the blaster looking like a parachute after it was stretched by an incompetent worker. There are knowledgeable, talented people out there; find them instead of the alternative.

To find out more about abrasive blasting, I contacted Bill Durbin of Broadway Blasting Services of Hanover, Pennsylvania (717-630-2260). I have been to his shop many times and have personal experience with the quality of his work. In our discussion, Durbin stressed the importance of choosing the correct material and using it properly. For most jobs, Durbin prefers one of three types of plastic media, depending on the substrate. Urea, acrylic, and melamine have different characteristics that make them suitable for different purposes. Plastic blasting media are ground material with sharp edges, Durbin explained, that literally drag the paint from a surface. If you have ever looked closely at a piece of steel that was blasted with sand at high pressure, you know that the surface resembled a piece of 80-grit sandpaper. In contrast, plastic deforms when it hits the surface, so it does not abrade like sand, nor does it endanger your life through silicosis, as does sand. By adjusting the air pressure, media flow, blasting distance, and angle of attack, the blasting professional has considerable control over the aggressiveness and speed of his or her work. Durbin compared blasting to the beam from a flashlight. Imagine the beam shining directly down on a flat surface at close range, and then gradually moving away at an angle. The blaster needs to be cognizant of these factors to be able to work safely and at a reasonable rate.

For corroded steel, Durbin prefers aluminum oxide, which he blasts in a large two-person cabinet. For cleaning metal without biting into the surface, Durbin likes glass beads. After any media-blasting operation on steel, Durbin recommends sanding it with 180-grit or a Scotch-Brite pad, wiping the surface with a wax and grease remover, and priming it as soon as possible to prevent oxidation.

For steel with slight surface rust, or a non-ferrous metal with mild corrosion, media blasting may be more than you need or can afford. Consider a metal-conditioning treatment. There are a number of phosphoric acid-based metal cleaning solutions available that will return metal to a pristine state and keep it that way for some time with a treatment of a metal conditioner.

Our bare metal handmade Ford Model T speedster body was rusty and needed to be cleaned prior to being painted. Because we use PPG paints, we cleaned the body using PPG's DX 579 Metal Cleaner and then applied DX 520 Metal Conditioner. DuPont and other companies have similar products. Following the conditioning treatment, we painted the body with PPG's DPLF Epoxy Primer.

For metals other than steel, consult the PPG website for the proper ratio of metal cleaner to water (www.ppg.com). For our steel body, we mixed the metal cleaner with water in a ratio of one part cleaner to two parts water in a squirt bottle. Working in one small area at a time, we sprayed the cleaner on a Scotch-Brite pad, wiped the metal, allowed the cleaner to sit for a few minutes, and scrubbed the surface with the Scotch-Brite pad. If you allow the material to dry on the surface, you will get poor results, so don't try this outside in the wind and sun. Rinse the area with water, wipe it with a towel, and then dry it with compressed air. Repeat if necessary to remove any residual rust. Following the cleaning process, apply the metal conditioner to a new Scotch-Brite pad, and working in small areas, wipe the surface, allow the conditioner to sit a couple of minutes, and then rinse and dry as you did with the cleaner. Wipe down the entire surface with a wax and grease remover and you are ready for primer.

## METAL STRAIGHTENING

After you have dealt with old surface finishes and corrosion, you will be ready, perhaps even eager, to move on to greater challenges. If the 10 pounds of Bondo you just stripped off the fender of your project car revealed some latent damage, you will now have the opportunity to develop and hone some new skills and to right the wrongs that others have hidden under body filler.

Whenever sheet metal is hammered, bent, damaged, or shaped, its grain structure is altered and the metal becomes increasingly stiffer and stronger. The tendency of metal to become stiffer through cold working is called *work-hardening*. Sometimes work-hardening is an intentional part of the forming processes that lead to an object's creation. Whether the end result is a bucket, a bedpan, or a priceless *object d'art*, work-hardening adds rigidity and helps the metal object keep its shape. Other times, work-hardening comes about by accident, like when your teenager backs the family car into the steel basketball goal set in concrete adjacent to your driveway, twice. The impact damage of the basketball goal changes the structure of the steel in your trunk lid, making the lid harder to straighten with each new episode.

Work-hardening endows metal with a memory that can work for or against you. In the case of most automobile panels, for example, collision damage may often be corrected with a few well-placed hammer blows. As you begin to massage the damage with a hammer and dolly block, the metal's memory helps the panels reassume their original shape. Whenever possible, therefore, try to recover the residual memory from your object's creation and use it to your advantage. While there are times when the memory of a piece of metal cannot be recovered because of massive damage, this is the exception rather than the rule.

You need these supplies to apply PPG's Metal Cleaner and Conditioner: rubber gloves, some Scotch-Brite pads, clean towels, a dust mask, wax and grease remover, clean water, and some squirt bottles to apply the solutions.

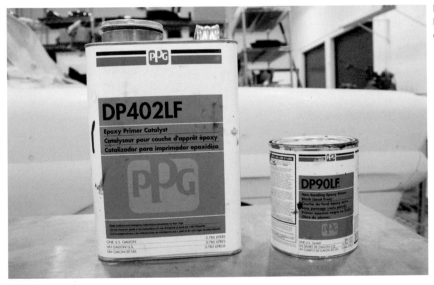

PPG recommends their DPLF Epoxy Primer for use on bare metal that has been cleaned with their cleaning and conditioning solutions.

The right side of our Model T speedster body has surface rust. The left side has been cleaned and treated with PPG's Metal Cleaner and Metal Conditioner.

To illustrate the concept of work-hardening through cold working, bend a piece of light-gauge steel vigorously back and forth by hand. The steel will heat up noticeably from the internal friction within the grain structure, and the steel will get progressively harder to bend. For the strongest and most tenacious readers, the metal will eventually tear along the bend because the grain structure can no longer accommodate further deformation. Repeated bending is just one form of work-hardening. As alluded to previously, intentional work-hardening may result from the initial die-stamping, folding, extruding, rolling, or beading operations that give an object its shape. Intentional work-hardening is generally your friend. Accidental, or unintentional, work-hardening, on the other hand, is a nuisance resulting from impact, vibration, clumsy repair efforts, and other causes. Now stiffer than ever, accidently work-hardened metal can be difficult to repair because it requires more force to move than the undamaged metal. Consequently, you might mistakenly stretch undamaged adjoining areas on your panel. I will cover heat-shrinking techniques for repairing work-hardened metal later in this chapter. For now, keep the phenomenon of work-hardening in mind as you think about the most stubborn dents on your project vehicle; you will not want to apply any more force than necessary to remove them.

To return bent metal to its previous undamaged condition, you must first avoid exacerbating the damage by hammering semirandomly in an attempt to will the metal back into shape. Such antics will likely lead to stretched metal, frustration, lifelong ridicule from family and friends, and overall despair. On the contrary, take a moment to look at the damage and consider what caused it. You do not need an exact narrative of the sordid details, only a general idea of what got pushed where. As Frank T. Sargent first proposed in *The Key to Metal Bumping*, the best way to repair damaged sheet metal is to remove the damage by working in reverse order from the way the damage was created. For example, imagine that a delinquent youth has dented your car door with a horseapple. Following Sargent's approach, you would reverse this dent by pushing with your hand, or perhaps gently hammering, on the back side of the panel starting from the outside perimeter of the dent, gradually working your way toward the middle, because this sequence reverses the order in which the metal was deformed. Hopefully, the memory of the metal would allow it to reassume its original shape. I don't know if fingerprints can be lifted from a horseapple, by the way.

Sargent's approach is a good starting point for straightening sheet metal, but there will be times when you either won't be able to discern the sequence of events that caused the damage you hope to fix, or you will know how the damage got there, but you'll be puzzled nevertheless as to how to reverse it. Perhaps the realization that sheet metal changes in a limited number of ways will make your repair attempts less intimidating. Metal that requires straightening will be damaged in one or more of the following ways: it will be bent, work-hardened, stretched and/or upset. As I mentioned in an earlier chapter, a bend is similar to a fold or crease in a piece of paper—think of two planes meeting along a single axis, the bend line. If a bend is slight, such as 10 to 20 degrees, the metal can be returned to its original state fairly easily by reversing the bend, and the metal is the same thickness that it was before. Technically speaking, a bend creates a stretch on the outside of the bend and an upset on the inside of the bend, but ignore those details for a moment and blithely read on so that the concept of a bend will be firmly planted in your mind. I believe that thinking of a bend as distinct from the more elusive concepts of shrinking and stretching simplifies metal straightening. Bent metal often behaves as if it were hinged—straighten the metal by pivoting it back into position along the bend line. This is the simplest form of repair, so appreciate it when it comes your way.

## TROUBLESHOOTING

Because metal is crystalline in structure, damage to one spot easily influences surrounding metal. Imagine the big kid on the playground crashing into the Red Rover line. Locked arm in arm, all of the kids fall higgledy-piggledy onto one another. Hopefully, none have separated shoulders. Grains of metal have a similar relationship to one another in that an impact in one area inevitably evokes a reaction somewhere else close by. Most deformation takes place along slip planes within individual grains of metal. Depending on how grains are oriented in relation to one another, how closely packed individual planes of atoms are within each grain, and how much space lies between parallel planes of closely packed atoms, slip, or deformation, can travel across grains.

As damage increases beyond a simple bend, more grains of metal are brought into play. We must think of the damage as occurring in an area or zone, rather than along a line. Stretched metal is longer, wider, and thinner than it was previously, whereas upset metal is shorter, narrower, and thicker. Imagine trying to return a piece of partially chewed gum to its original pristine state. With the right tools and application of force, it could probably be done, but you would anticipate some challenges, especially if the gum had been stuck on the underside of your chair for a while. Straightening sheet metal is a little like this, but it's easier and sheet metal usually is not sticky, though it might be if it has been hit by a horseapple.

Fortunately, the worst obstacle you will encounter when learning how to straighten sheet metal is greatly diminished once you've been warned about it. That obstacle is this: your natural inclinations will lead you astray until you've gained some experience. Let me give you an example to illustrate this point. Imagine that you have a flat piece of steel with a small raised area created by the impact of a small object against the back side of the panel. Your natural inclination, of course, is to lay the metal down on an anvil and smack the high spot with a hammer repeatedly. At first this will work, you'll feel confident, and your early success will spur you on. As you get the panel almost flat, however, the metal will seem to spring back. You'll grit your teeth and press on, hammering harder in hopes of flattening the dent. If your work surface or your hammer is softer than the metal, you might be able to flatten the dent, but if your hammer face is steel, inevitably your panel will start to bow as the metal is squeezed, thinned, and stretched between the hammer and the anvil face. The site of your hammering will become tight and as hard as armor plate. The bulge resulting from your hammering may be worse than the original damage you hoped to repair! If you push on the bulge with your hand in an attempt to make it flat, the metal will pop back and forth with a *ka-thunk*. This situation can be fixed, and I will offer a remedy later in this chapter, but the important point to remember is that this scenario will repeat itself every time you

*aggressively* hammer metal against a metal surface. The process of squeezing the metal against the anvil has work-hardened the metal, making it very stiff, and it has changed the metal's physical size. The hammered area is thinner, but takes up more space in the 360 degrees surrounding the worked area. As a result, the hammered area creates tension between this part of the panel and the surrounding unchanged metal and will *ka-thunk* inharmoniously every time you push on the panel. Sometimes the metal is severely stretched and work-hardened to such an extent that it does not flex, nor make any noises. The metal simply bulges excessively in comparison to the original contour. If you encounter metal that some prior unfortunate soul has attempted to straighten, but abandoned, you will almost certainly have stretched metal such as the aforementioned bulge to resolve. The solution to repairing stretched metal is to shrink it, which I will cover in a moment.

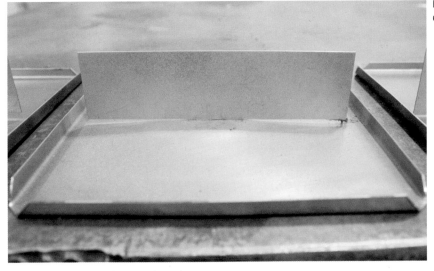

Here is a high spot much like you might need to straighten on an automotive restoration project.

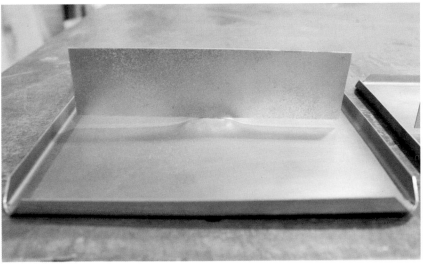

If you try to flatten a high spot by vigorously hammering it against a dolly, you'll end up with a smaller, harder high spot like this. The metal has been smashed against a dolly. In the process, extra length has been added because the metal is thinner but longer in the hammered area.

Heating sheet metal causes distortion by creating an upset, or shrink, adjacent to the heat zone on all sides. When the metal cools, there is not enough tension to pull the metal back to its original flat, pristine state. The heated spot is shorter and acts like a trampoline bounded by unspoiled metal.

The way to fix overshrunk metal is to stretch it. Hammering this piece against the table in the overshrunk area released all of the tension so it would lie flat again.

Problems with shrunk or upset metal are less common than problems with stretched metal for at least two obvious reasons. First, sheet metal stretches much easier than it upsets, and second, the factors required to induce shrinking are less likely to occur by chance than the factors for stretching. To be upset successfully, sheet metal must be held tightly and have force applied so that the grains of metal butt into and elide with one another without the metal bending excessively. At an atomic level, planes of atoms must shear over one another. This situation can happen in an automobile collision, for example, but any shift in the alignment of the metal in relation to the force applied will usually result in something bending and stretching rather than the creation of an upset other than in a very narrow strip on the compressed side of a sharp crease.

Shrinking can also result from the application of heat to sheet metal. Heating a panel with a torch or by welding can cause distortion as a result of shrinking. Grains expand when heated, but their lateral movement is confined by cooler metal surrounding the heated spot. The restrained expansion of the hot metal creates an upset where the hot metal meets the cooler surrounding metal. After the metal cools, the formerly heated portion takes up slightly less space than before due to the upsetting. As a result, the shrunk section of metal pulls on the surrounding metal, resulting in distortion. Unfortunately, unless there are obvious visual indicators like welds or steel colored by heat, upset or shrunk metal is difficult to pinpoint by sight. Stretched metal, in contrast, is usually easy to identify by sight; it usually bulges in or out. Although there is nothing to broadcast its location, upset metal may often be found adjacent to damaged metal. Whether or not you can identify the upset metal's location may be a moot point, however, because you will be forced to resolve the upset during the normal dent removal process anyway. Stretching out low areas and heat shrinking stretched areas will naturally resolve any upsets lurking nearby. Now that you have a better idea of the kinds of damage you are likely to encounter, let's get into some hands-on exercises.

For the first demonstration, I will modify an antique automobile fender for educational purposes with a large hammer. The best way to learn to repair sheet metal damage, of course, is to spend some time getting acquainted with the material. An old automobile fender is an ideal candidate because of its manageable size, workable thickness, and built-in memory left over from its original stamping. Prepare your specimen with a large smooth-faced hammer, cannonball, or steel-toed boot. Small-faced blacksmith, machinist, or ball-peen hammers will leave deep scars that will complicate repair efforts. Save those for future experiments. If you do not have a lot of experience with dent repair, your goal will be to create a large impression or set of impressions that you can practice removing by working in the reverse order that the damage was created, using the memory of the metal to help spring the panel back into shape.

When approaching a dent to repair it, don't be afraid to make creative tool choices. On large dents such as the one I am recommending that you produce here, it makes perfect sense to select a large rubber mallet, a dolly, or perhaps that old shot left over from your high school track and field days, not because these items are heavy, but because their large faces spread the force of your blows, thereby reducing the chance that you will accidently damage the work with nasty tool marks. Beware that heavy tools are more likely to stretch metal, however. Furthermore, the curvature of the tool face should ideally match or at least resemble the curvature of the panel. Sometimes the palm of your hand is the best tool. If possible, lay the fender down on a workbench with a shot bag behind the damaged area. Starting at the outer perimeter of the dent, work your way around the outside of the depression from the back, working gradually toward the center with each pass. For dent repair on the workbench, the shot bag outshines the sand bag in my opinion, because it supports the work like malleable clay, which helps you control the strength of your hammer blows and makes you less likely to stretch the metal than if the panel were unsupported.

The best way to learn to repair dented fenders is to repair a dented fender. Use a large smooth-faced tool to create a dent if you don't have one already.

Unfortunately, I have been a little overzealous in creating my dent on the sample piece, so I will not be able to rely solely on the memory of the metal to bring the panel back into shape. The wrinkles in the metal indicate that some more work is needed. As you progress in dent repair, you must adapt your techniques to the stubbornness of your metal. After my first round of straightening with a dolly, some creases require additional attention with a bumping hammer on the inside of the fender while I support the area with a small shot bag on the outside of the panel. Although very small low areas may be raised by hitting them over a crowned dolly with a slapping spoon, keep in mind that this technique can lead to too much crown through stretching. Note, too, that the beautiful smoothness the slapping spoon creates can be deceiving. Do not be seduced by smoothness if the panel you are working on is not the proper contour. Go over the area lightly with a large flexible sanding block to reveal highs and lows, or use the old body man's trick of swiping a cotton cloth over the panel with your hand flat against the surface. The cloth eliminates false feedback from oil on your skin or sanding residue on the panel, thereby giving you a true sense of the panel's contour. Instead of hammering or slapping on top of the dolly, placing the dolly under the low spot and hammering next to it—the so-called hammer-off technique—is a safer route. I am convinced that each piece of metal responds differently to hammering, so it is essential to try several techniques to find what works on that piece rather than assuming you can always use the same routine.

The dent in the sample fender is now approximately 80 percent finished. The traditional method of the last phase of dent repair is known as *picking and filing*. Auto body technicians in days gone by would run a vixen file over the surface to reveal high and low spots. Then they would bump or pick low spots up and hammer high spots down. This method is not downright evil, but it has a couple of shortcomings. First, a vixen file is easily abused, which can lead to thin metal. Second, by taking the *picking* part of the process too literally, a worker who uses a pick hammer to raise every low spot will end up with a hopelessly bumpy panel. Often, the worker in the second case is the same one who does not know when to stop filing, so he or she promptly files the tops off the bumps and the panel looks as though it was shot with birdshot from a shotgun, a situation that is surely worse than the original dent.

We've lightly gone over the back of this dent with a sanding block so that the elevation changes are more noticeable. This is a classic case of trying to repair the dent in the reverse order it was put in. That means starting at the edges with a big-faced tool such as the general purpose dolly.

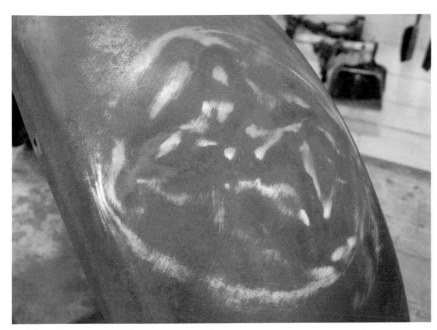

After the first round of pounding from the back with a dolly, the dent is better, but the surface is still a little rough.

Next we'll use the plastic teardrop mallet against the shot bag to try to raise some of the low spots.

The modified version of picking and filing that I recommend is to go over the surface with a sanding block instead of a vixen file. The purpose of sanding is to gather information about the surface of the panel. You are *not* trying to sand the panel smooth. To make the panel's topography more obvious, you can lightly fog the surface with a sandable self-etching primer. I like the Dupli-Color self-etching primer available at O'Reilly Auto Parts stores in our area. I have not analyzed the ingredients of this product to determine what makes it so sandable, but it works splendidly as a guide coat when lightly applied. I have tried one or two other less expensive so-called sandable spray can primers that were miserable failures. Any enamel spray paint will be a gummy sandpaper-clogging nightmare. If you go the Dupli-Color route, remember that you are simply misting the area to be worked. You are not camouflaging a duck blind. If you put the paint on too thickly you will be adding surface thickness and end up wasting a lot of time sanding primer. Consider, too, the compatibility of the self-etching primer with future paint treatments. If you will be painting the part, you will be sanding it anyway prior to paint, but you do not want to run into problems if your top coat prohibits the presence of self-etching primer.

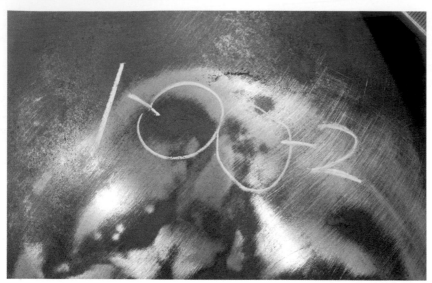

Here is a close-up of two kinds of damage. I would call dent No. 1 something that should only be hit from the back into a shot bag, whereas dent No. 2 is subtle enough that hitting it with a spoon on a dolly might be the best course of action. The danger lies in bashing deep dents like No. 1 on top of the dolly and severely stretching the metal.

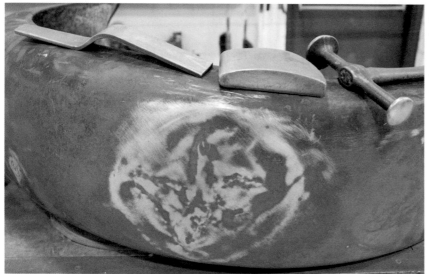

Oddly enough, sometimes dents look worse before they look better. After some work with the teardrop mallet, there are fewer low spots, though the sanding block reveals a surface that still needs some work. A situation like this is where the hand-held shot bag really shines. You can work the inside of the dent with a crowned hammer, and support the exterior of the fender with the shot bag held in your other hand.

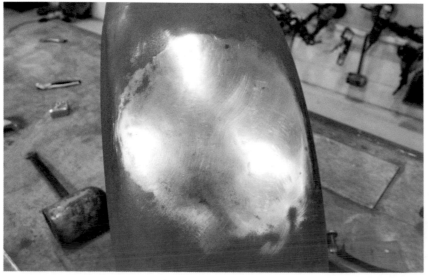

Metal straightening can be slow work. It requires patience and accuracy. Resolve one area at a time, but eventually it will all come together. In the absence of a shot bag, hammer off dolly is the best technique; push on a low spot with the dolly and hammer next to it to trade a high spot for a low spot.

Whether or not you use a self-etching primer guide coat to help you straighten your panel, the overriding strategy at work is to flatten the metal by lowering high spots and raising low spots. Rather than relying heavily on the pick hammer to raise the low areas, use a hammer with as large a face as you can get away with to spread the force of your blows. Try to match its face to the contour of the panel you are trying to straighten. Curved hammer faces are for hitting on the back side of crowned panels, while flat faces are used for hitting on the exterior side of a crown. If the hammer face you have chosen is too large or does not have enough crown, it will hit the panel around the dent prematurely and leave hammer marks. Furthermore, your hammer face must always be flush to the panel when it makes contact; otherwise, hammer marks will result. You will make the most progress and cause the least amount of damage by placing a dolly under the low spots and hitting the high spots or, alternatively, working from the back of the panel hitting against a piece of wood, cardboard, or a shot bag. Most dent repair is a leveling process through which displaced metal is pushed back where it belongs.

## HEAT-SHRINKING

By working in the manner I have just described, you will not have much trouble removing a simple dent. Let's increase the challenge, therefore, by severely stretching the metal past the point that its memory can help recover the shape. Lay your fender down over a shot bag and smack it once soundly with the ball end of a ball-peen hammer. Hopefully, you'll end up with a tumorous protuberance on your fender resembling the beauty on the demonstration piece.

At this point, you may either succumb to the urge to hammer the large lump flat or not. If you do, you will make the problem worse, but by doing so you will only enhance the educational value of this exercise. The only way to solve this stretched metal problem, besides making a new piece, is to shrink the stretched metal. There are several ways to shrink stretched metal. First I will show you the oxyacetylene torch.

As a lifelong student of metal working and dent repair, you are likely to encounter several variations of torch shrinking. I'm sure they all work for somebody somewhere. By all means try them all and practice what works for you.

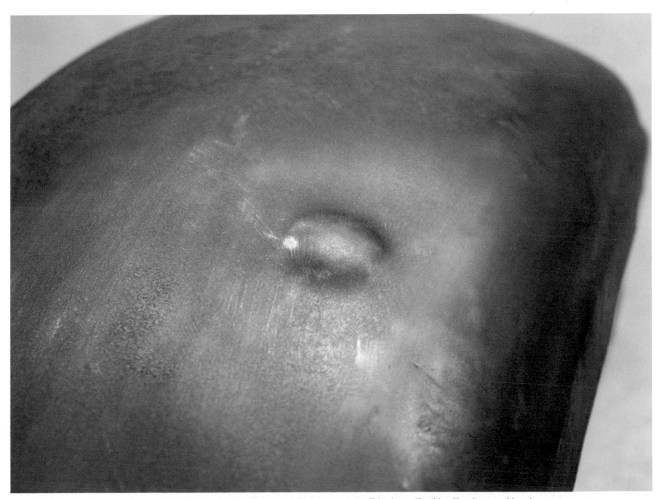

Hopefully, this eyesore is worse than any naturally occurring dent that you are likely to encounter. Fixing lumps like this will make everything else seem easy.

The following is the method that I prescribe for heat-shrinking steel. Find a dolly with either the same or slightly less curvature than the panel you would like to shrink. Place a small dinging hammer and torch stand near your work so that you will be able swiftly and safely to rack the torch and grab the hammer. Adjust an oxyacetylene torch just as you would for welding and heat a spot about the size of a dime until it is dull red. Typically the heated spot will rise up as it expands. Quickly place the dolly behind the heated spot with one hand and place the torch on the torch stand with the other hand. Grab the hammer with the torch hand and smack the heated spot once, supported behind by the dolly. Use a blow as you would use to drive a tack. You will hear a dull click rather than a high-pitched metallic clink as you hammer the heated metal down against the dolly. A metallic clink is a dead giveaway that you are hitting the metal too hard. The first blow will flatten the top of your heated lump so that it resembles a tiny mesa. After the first blow, gently tap around the perimeter of the mesa to smooth and flatten the lump. Do not feel compelled to remedy the situation completely on the first attempt. Often more work is needed to smooth a shrunk area completely.

Torch shrinking sounds easy and it is, but I have watched it done improperly enough times to know that a few cautions are in order. The margin for error between *just right* and *way wrong* is slim. First, you *must* hit the heated spot while it still has color. Time is of the essence, so if you fumble with your tools, simply wait a moment, reheat the metal, and try

To torch shrink successfully, heat a spot about the size of a dime red hot. The illustrated piece is a little hotter than necessary, but I wanted to make sure that it would be visible in the photograph.

Hit the hot spot about as hard as you would drive a tack. Support the back side with a dolly having slightly less crown than the panel.

224

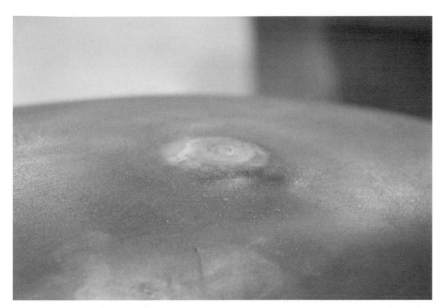

After the first hammer blow, the lump should resemble a tiny mesa.

Following the first blow, continue hammering around the rim of the mesa to smooth the dent further. If the dent seems springy and refuses to shrink further, more applications of heat may be necessary.

again. Second, some people have a tendency to try to touch the hammer to the work once before swinging it, presumably to improve their aim. This takes too much time. You must pick up the hammer and swing it down abruptly. Third, do not hit the metal too hard or you will stretch it by smashing it against the dolly. I would define *too hard* as the force you might normally use to drive an 8d nail into a piece of pine. Fourth, do not get greedy and try to heat a larger spot than perhaps a nickel. You will be unable to trap the heated spot against the dolly within the needed boundary of cooler metal. Instead, proceed in small, controllable steps. Keep in mind that the first hammer blow does the work of shrinking and subsequent blows simply smooth the area. When you first heat the panel, the metal rises up due to expansion. It bulges

up because that is the path of least resistance. Your first blow on the peak of the bulge forces the hot metal to upset around the sides of the bulge. The hammer face applies force on top, the cooler metal surrounding the hot spot resists the force around the perimeter, and the dolly keeps the metal from simply being pushed down when you hammer it. Allow the metal to air cool and assess your progress by running your hand over the area. The metal must be cool enough to touch before you will know how successful your shrinking attempt was. Do not be surprised to find that you need to repeat the process a few times in the same general area. A lump that you can cover in the palm of your hand, for example, might require one shrink in the middle and perhaps five additional shrinks around the edge.

225

To perfect your heat-shrinking technique, set up a test panel such as this and practice. You can also leave the torch in the stand and move the panel with one hand in and out of the heat while you hammer it with the other hand against the table. The first method is better training for dealing with a panel on a car, however.

The torch shrinking method I have just described is the most foolproof version I have found. If any of the factors are not just right it will not work, but once you have mastered it, it will always work. Torch shrinking will bring you peace of mind because it adds a little control and predictability to dent repair. Fortunately, practice panels are easy to set up, so I encourage you to be proactive in your education. Don't wait for the perfectly stretched panel to come along or I promise it will come in the form of something irreplaceable and at an inconvenient time. Make your nightmarish panel yourself. For laboratory experiments such as this I recommend bending a small flange along each side of your test panel, but leave the corners open to allow the panel to twist and distort in case you overshrink the panel. The flanges will stiffen the metal and make it behave more like a panel that has been stamped or formed into a shape you need to repair on a car. Whether you are straightening a breadbox, a Radio Flyer wagon, or a Ferrari, metal will usually behave differently once it has been formed and damaged than when it is pristine and flat because of the stiffness of the surrounding metal induced by work-hardening. Put a small flange on each side of your test piece, lay it flange-side down on a shot bag, and hit it on the back with a ball-peen hammer. Turn the metal over, clamp one end

to the edge of a work bench to stabilize the panel and practice your technique.

## Variations

Other variations of torch shrinking that I have come across recommend quenching with water in addition to the steps I just described, or they recommend against supporting the hot spot with a dolly. I do not recommend quenching for your first attempts because quenching seems to intensify the shrinking effect and lessens your control, especially on a panel that is flat or nearly flat. After you have mastered torch shrinking without quenching, set up a flat panel and try quenching. In addition, I have tried torch shrinking without using the dolly and it seems less effective than shrinking with the dolly, in my opinion. When shrinking without the dolly, the hot metal has a tendency to push down below level, rather than upset into itself as it does with the dolly behind it. I presume some shrinking nevertheless takes place with your first hammer blow, but my overwhelming impression is that after the initial hit the metal feels like lasagna al dente.

Once you have gained some confidence torch shrinking on test pieces, you will no doubt try torch shrinking on a large steel panel with a gentle crown, such as a car door. You will be merrily shrinking, perhaps even quenching your shrinks to speed the process along, when all of a sudden the panel will become concave with a sickening *ka-thunk*. You have overshrunk your panel. You might also run into this shrunk metal condition adjacent to a weld bead. In either case, the solution is to stretch the panel back out. One extremely low-tech method of remedying this situation is to lay the panel face-down on a piece of cardboard on a workbench. Position a body spoon over the low spot and hammer the spoon. The force will be spread out enough that the panel will not be damaged and the overshrunk condition will be eliminated.

If you are anxious to tackle overshrunk metal, prepare a test panel as I just described for torch shrinking. Do several torch shrinks and quench each shrink promptly after you hammer it. Hopefully, the panel will buckle hideously. Lay the panel on a workbench and admire your work. If the panel refuses to lie flat, you have done a good job. To remove the distortion, turn the panel over so that it is supported only by its flanges. Hammer the back side gently with a plastic or rawhide mallet. You can also try hammering the panel against a flat dolly, a piece of wood, or a piece of linoleum over a hard surface. It may take a little experimentation to find just the right combination of tools to give the degree of stretching you need, but it is possible to get this panel flat again without an oil can by stretching. Shrinking created the problem, so stretching will take it away. Torch shrinking as I have described it barely scratches the surface of what is possible with heating and cooling steel to shrink it, but if you succeed with the experiment I have just proposed, you will have made great strides toward being able to predict when and how much to shrink.

## SHRINKING WITH RESISTANCE WELDERS, STUD GUNS, AND SHRINKING DISCS

Heat-shrinking with the oxyacetylene torch is one of the most useful metal-straightening techniques you can learn, but it is not the only way to shrink steel. Sometimes you need to shrink a very small spot and would like to do so without resorting to the torch. The resistance welder, the stud gun, and the shrinking disc are other useful weapons in your arsenal.

Resistance welders and stud guns are primarily used for welding applications, of course, but they can be effective heat-shrinking tools as well. Both tools are quite expensive new, so I would not recommend buying either of them solely

A low-tech but effective way of stretching overshrunk metal is to hammer a spoon against it from the back side. The spoon will spread the force of the blow and leave no trace. Put a piece of cardboard underneath so the metal can give a little when struck.

for their utility as metal-shrinking devices, but if you have the opportunity to borrow one or buy a used one cheaply, it might be worth your while to do so. Resistance welders operate by sending electrical current through two electrodes that trap the metal to be welded. The resistance offered by the sheet metal pinched between the electrodes heats up the metal enough to weld it at the pinch point. When welding, clamp the work firmly between the electrodes and hold the on switch for several seconds. Because they rely on good electrical contact for their operation, resistance welders can be a little finicky to set up for welding, however. The electrodes have replaceable copper tips that should be a certain shape, and they must meet the panel to be welded squarely and with a specific pressure. For heat-shrinking, thankfully, none of the exactitude needed for welding matters. For problems such as the one presented by the flabby Fairlane trunk lid later in this chapter, or for small, localized shrinks, clamp your stretched metal between the electrodes. Engage the on switch for a count of *one thousand one*. If you hold the switch too long, the electrodes will burn a hole through your metal, so proceed with caution. Heat the metal just enough to create a very small upset, which will appear as a blue spot. Expect to carry out several small shrinks because the effect is very subtle. Unfortunately, each tiny heated spot will also be marred slightly by the electrodes, so you may decide that you cannot live with small pockmarks on your panels. If your metal will be sanded, the blemishes will most likely be removed though.

Similar to resistance welders, stud guns are capable of creating small upsets through heating. Stud guns are typically outfitted with a tip for welding a small metal pin into a low spot on a steel panel. A slide hammer is then used to grip the pin or pins and thereby apply pulling pressure to a hard-to-reach dent. When the welding tip is changed for a heating tip, however, the stud gun is ready for heat-shrinking. To use the stud gun for shrinking, support a steel dolly behind the area to be shrunk and press the spring-loaded heating tip firmly against the panel. Squeeze the trigger on the gun for a count of *one thousand one, one thousand two* and remove the tool. Carry out additional shrinks if needed. With either the resistance welder or the stud gun, make sure the areas you intend to shrink, as well as the tools' electrode tips, are silvery clean. Not only will these machines not work properly if rust, paint, and dirt are present, they will arc angrily and erode the metal at the site being heated.

The resistance welder can also be an effective shrinking tool.

The conventional stud gun is on the left, and the stud gun with a heating tip is shown on the right.

## STRAIGHTENING A DOOR

In an effort to apply several of the metal straightening techniques described in this chapter, I have rescued the perfect candidate for repair from our stockpile of unloved metal. I shook my head in disbelief when I first ran my hand across the grotesquely stretched skin of this Ford Model T door. The sheet metal bulges drastically in and out in various places, and to make matters worse, or I should say better in terms of the educational opportunity provided, there is very little access to the door from the back side. Limited access usually necessitates creative problem solving, so repairing this door should be a fruitful exercise.

This old door has four general areas of damage that are sure to require all kinds of trickery to repair.

Lightly sanding the back side of any dent will help to reveal high and low spots.

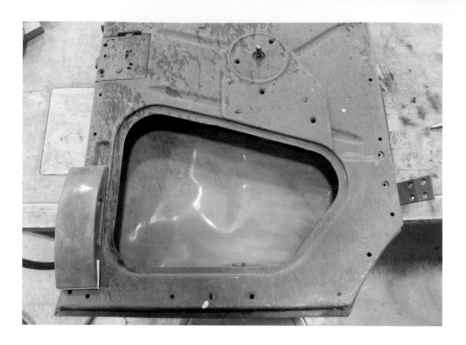

Always start your dent repair with the least invasive means possible. In the present case, many of the large displaced expanses of metal are easily pushed out by hand, but the metal is so stretched that the dents return the moment I remove my hand from them. Recalcitrant dents like these require shrinking.

Having made no progress working by hand, I find an assistant and address the large rigid vertical crease marked *1* on the door. By wedging a large spoon behind the inner door structure and pushing up on the crease from the back side, we perform several shrinks with the stud gun to the crest of the crease using the gun's heating tip. In places where the spoon will not reach, we apply pressure to the underside of the panel by wedging a dolly inside the door on top of a small block of wood. Our shrinking with the stud gun works well, and the creases we were concerned about do not return. Although the metal will need some more smoothing eventually, our main goal now is to return the worst areas to the proper contour. The shrinking involved thus far has already tightened up some of the floppiness in the skin overall.

We will wedge this spoon inside the door to provide support for the stud gun's heating tip. Without something behind it, the heating tip will push the hot metal below level.

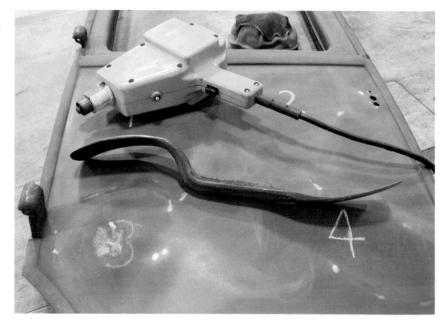

In the upper part of the door where access is restricted, we will squeeze this dolly behind the area we want to shrink. A block of wood will help hold the dolly tight against the panel.

Laying the door face-down on a piece of cardboard on the workbench, we sneak up on the deepest remaining dents by working from the outer perimeter toward the deepest part of the dent. Often bashing directly on the peak of a crease will further upset an already work-hardened area, but if you work from the outside in you can straighten the area without adding new damage. In the tight inaccessible areas behind the door support structure, we bump out dents with a curved spoon used as a driving tool. We gauge our progress on the door by brushing over the surface with a sanding block to reveal high and low

spots. As areas needing attention are identified, we level them with the hammer and dolly using predominantly the hammer-off technique.

A crease near the area marked *5* on the door provides an opportunity to shrink with the resistance welder. The latter is not quite as maneuverable as the stud gun because you must have a set of electrode arms to reach around obstacles, but its ability to shrink is comparable to the stud gun once you have reached the target. We close down the electrodes on the metal, hold the trigger for a count of *one thousand one* and the metal shrinks due to the tiny upset that occurs.

Hammering a flat spoon against the back side of the door allows us to raise some of the low areas.

Spoons can be good driving tools for reaching into areas that are hard to access, such as the corner of this door.

We will straighten area *4* using a hammer and dolly and the hammer-off technique.

The crease in area *5* gives us the opportunity to shrink using the resistance welder. A momentary heating of the panel between the electrodes provides a tiny upset.

Just below area 5 on the door is a dent that is inaccessible from the back side because it is covered by the door's inner structure. Situations such as this call for the stud gun to be used in its conventional mode as a welder. A small steel pin is inserted into the welding tip, and the tip is pressed against the panel. The trigger is squeezed for a couple of seconds, and the metal pin is welded to the panel. A special friction-operated tip is attached to a slide hammer, and the pin is gently pulled to raise the metal to which it is attached. The smoothest results are achieved by attaching several pins as close together as possible rather than applying more force in only a few spots. We attach two pins to this dent.

A panel with restricted access tests your patience and resourcefulness. With our door's inner structure limiting our access to the back side of some dents near the edge of the door, we call into service one of the most oddly shaped and seldom-used prying tools in our shop, a giant hook with a 90 degree bend at its tip. We maneuver the tool into position, get a general idea of the location the tip according to where we can cause a bulge in the outer door skin by rocking the tool, and then gently tap around the bulge with a hammer. A few sweeps with the sanding block allows us to check our progress overall.

The stud gun is the unrivaled champion for removing inaccessible dents. The door's inner support structure prevents access to this area from behind, so we will attach a stud or two in place and pull from the top side.

A specialized slide hammer is used to pull on studs from a stud gun. In this case, the stud is held tightly by a knurled wheel that grips the stud when tension is applied to the hammer handle.

Keeping continuous tension against the stud and hammering around the stud will give you a different pulling effect than just yanking on the stud alone. The former technique offers better control, I think.

For the final phase of work on the Model T door, we apply a light coat of self-etching primer as a guide coat. Some areas near the top of door are inaccessible without cutting away some of the door's inner structure. At times removing structure is inevitable and makes the work to the outer skin much easier. With the Model T door we do not want to take such drastic measures. If you must separate an outer panel from its structure, make sure that the latter is stable enough that it will not change when the outer skin is removed. Add braces if necessary so that you do not have to try later to recover the lost original shape. We finish our repairs by leveling the highs and lows with a hammer and dolly, raising subtle lows with a slapping spoon on a dolly, and sanding intermittently to gauge our progress. Ironically, a project like this looks worse for a brief time before it gets better. The even layer of rust that previously coated the door disguised many imperfections. As the difference between the various high and low areas narrows, the door takes on a splotchy appearance. When the dark splotches seem mostly transparent, your panel is close enough to flat to be sanded with 180-grit paper. If no rust is present, you may prime the panel.

## THE SHRINKING DISC

Another popular tool for shrinking metal is the shrinking disc. Derisively called the *shrieking disc* by my colleagues at work, the shrinking disc is not a tool to use during late night work sessions within a few miles of sleeping humans. Nevertheless, for people who are not comfortable with the oxyacetylene torch, the shrinking disc is a low-cost way to dabble in shrinking. If you are comfortable with torch shrinking, however, you will most likely find the shrinking disc cumbersome and slow-acting.

The original shrinking disc was developed years ago by Scott Knight of Scott's Hammer Works in City of Industry, California. The corrugated stainless-steel disc generates friction when the tool is rotated at high speed against the work piece. The user applies the tool to the work, sweeps it over the surface until the offending stretched area bulges from heat expansion, and then quenches the hot area to shrink the metal. The disc illustrated here was purchased many years ago from Sunchaser Tools. Most recent iterations of the shrinking disc are based on Wray Schelin's design. His disc is flat, not corrugated, and has an upturned rim for safety (http://stores.ebay.com/pro-shaper-tools-and-videos). The rim both stiffens the disc and makes it less like a spinning saw blade than the original version. Smaller discs are available for 4½-inch grinders. The original corrugated version of the disc needs to be well supported with a hard rubber or fiber backing disc or it is prone to cracking. We used both supports for maximum safety.

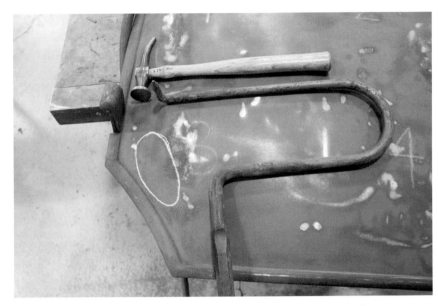

I won't conceal my delight at the opportunity to use this wacky prying tool to reach deeply inside the door and push on a dent from the back. Twisting the handle to the side applies leverage to the tool's tip.

Overall, the door looks pretty good. A little more hammer and dolly work will get it right on, but it has served its purpose as a demonstration piece.

Do not expect a shrinking disc to be a panacea for every metal-straightening problem. You must have some aptitude for straightening metal with hand tools before relying on the shrinking disc or you will be disappointed. The disc is not going to remove an inch-deep gouge in a panel that you have made no attempt to straighten, for example. It might, however, resolve that last 2 percent of damage that might otherwise confound you. Sometimes the final 2 percent of a repair requires more artistry and finesse that all of the previous work combined.

For the demonstration piece, I chose a trunk lid on a 1957 Ford Fairlane that was a little soft. At some point in the past, the leading edge of the trunk lid got caught on the adjacent drip rail and was bent slightly. The trunk lid was straightened as well as a person could reasonably expect. Nevertheless, the episode left the metal skin over the trunk lid less taut than it should have been because the inner support structure had been changed by the trunk lid mishap. The shrinking disc works on subtle shrinks such as we need to tighten up the trunk lid on the Fairlane, so we ground off the paint covering the soft spot, played the shrinking disc back and forth over the area until the soft spot swelled with tinges of straw color, then quickly quenched the area. After

two applications of the disc, the problem area became firm. Using the disc is that simple. Clean the stretched area until it is silver before using the disc so that you'll clearly be able to see any color change. Move the spinning disc over the area until you see a color change in the metal, then remove the disc and have an assistant quench the area with a wet rag. The water should sizzle when you apply it. Repeat the process until the metal has shrunk adequately.

Flabby stretched panels like the Fairlane trunk lid can be repaired with a torch, but the drama involved can be unnerving if you haven't done it before and don't know what to expect. To attempt this kind of subtle repair with a torch, have an assistant push up on the underside of the soft spot with a dolly while you gently play the torch over the top side. Before you see any color change in the metal, quench it. If nothing happens, heat the metal again but a little longer. I have used this method to tighten up flaccid sections of automobile panels before, but the process is tricky because you cannot predict how much to heat the panel. Once the color starts to change, the temperature is rising rapidly, so it is easy to overheat the panel. Also, the results of overheating a relatively flat panel will be dramatic, both to the panel and to your self-confidence. If you end up overshrinking the

This is the old-style shrinking disk and an archaic body grinder with a fiber and a rubber backing disk. The old style disk needs a lot of support to keep it from eventually cracking.

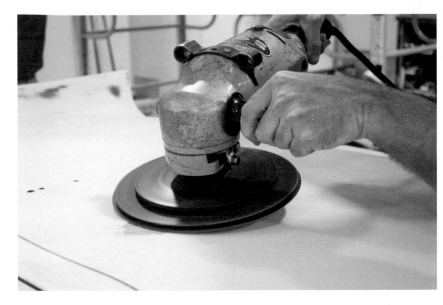

We play the shrinking disk over the soft spot on the Fairlane trunk lid until we see some color change in the metal. Presumably the heat generates an upset in the adjoining metal from expansion.

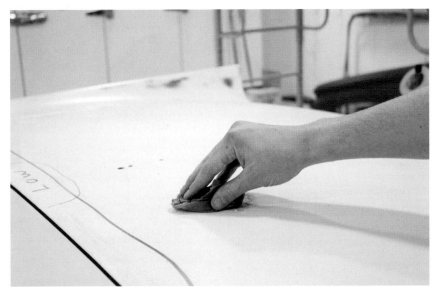

After each application of the disk, we quench the hot area with a wet paper towel. I assume that quenching prevents the upset created during heating from reversing. In other words, you get as much out of the upset as possible.

metal, the panel will suck inward alarmingly. Don't panic. Stretch it back out, just as with the sample panel earlier in the chapter. The Fairlane trunk lid could have been repaired with the resistance welder and the stud gun with the heating tip installed as well.

## WELD GRINDING AND SURFACE FINISHING

As a teacher of sheet metal restoration and panel fabrication, I have come to the conclusion that weld grinding and surface finishing are inherently difficult to learn. They must be, because I have witnessed such a large volume of nice work ruined through improper grinding. To finish sheet metal repairs so that they are indistinguishable as repairs and to weld together multiple panels without leaving a trace of evidence requires effective grinding. Furthermore, the preparation

of surfaces for paint or polishing relies on smoothing those surfaces by successive degrees.

The key principle to grinding welds is to grind only the weld bead. Too often, students attempt to compensate for uneven metal along a weld bead by grinding everything in hopes of achieving a flat panel. This method might work if the disparity between the panels is very slight and the metal is thick, but usually this approach is a recipe for disaster where automotive sheet metal is concerned. The high side of the bead bears the brunt of the grinding and ends up tissue thin. The only solution to thin, overground metal is to cut it out and patch it. Different welding methods leave weld beads of different characteristics, but in each case the purpose of grinding is to remove excess metal along the bead so that the weld is indistinguishable from the surrounding metal.

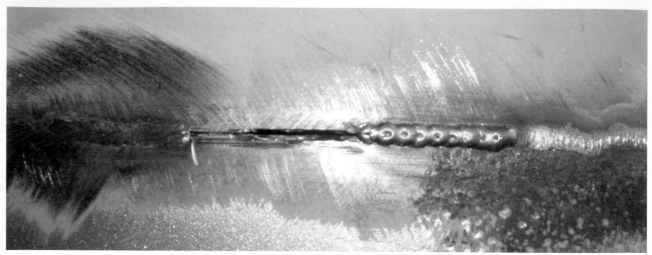

The partially repaired Triumph TR3 fender, or *wing*, in British terminology, shows an example of a ground MIG bead on the left, a cut made with a cut-off wheel at center to prevent binding between the panels, and an unground MIG weld at right.

There is no reason, therefore, to grind anything besides the bead. A TIG or torch-weld with a small bead will need very little grinding.

Your approach to finishing a weld bead will be determined by the bead. MIG welders typically deposit the most filler metal, so I will deal with the MIG bead first. I like to grind the highest part of a MIG weld bead down with a 3M weld-grinding wheel mounted in a pneumatic cut-off tool. Weld grinding wheels look just like cut-off wheels, but they are about ¼ inch wide. Whenever possible, keep them perpendicular to the bead as you grind or they will seek out the area on either side of the bead that does not need to be ground. An electric hand-held grinder with a 4½-inch disc works well also—almost too well—for the first phase of grinding. You will know when the bead is almost level because you will start to graze the area adjacent to the bead. When you get to that point, switch to another tool. A hand-held pneumatic grinder with a 5-inch 36-grit disc works well for taking a bead a little closer to flat. Again, try to grind across the bead, rather than along the bead, or the edge of the grinding disc will find its way next to the bead, where it will grind a trough in the metal. Use as much of the abrasive's flat surface as possible rather than tilting the tool up and using only the edge of the disc. Tap up low spots and bring down high spots with a hammer and dolly along the bead as you go or you will overgrind the metal for certain. Run either a sanding block or a vixen file over the surface at crisscrossing angles to check for flatness. The high spots will be shiny and the lows will be untouched. When the panel looks 99 percent flat, move to a soft abrasive backing pad with 80 grit, followed by 150 grit, and finish with a palm or random-orbit sander with 220 grit to be ready for primer. The classic novice approach is to begin with several costly 3-inch Roloc discs. These discs are great for working on small areas, but

they are not efficient for grinding large beads. In addition, their small tool face is less than ideal for achieving flatness; the situation is analogous to hammering a panel flat with a pick hammer. The novice spends a fortune on abrasives, takes forever, and ends up with a bumpy panel. When executed properly, the grinding of a MIG weld should not be a major production. Use as aggressive a tool as possible, but only as long as necessary to avoid overgrinding.

For TIG and torch-welds, start with a 36-grit 5-inch disc, a file, or perhaps the soft pad with 80 grit if the bead is not very proud. Finish with a palm or random orbit sander just as you would from this point with the MIG bead. If the metal will be left bare or polished, you will need to continue sanding with successively finer grits of paper: 180/220 to 320/400 to 600. From 600 you will be able to move to a buffing wheel if a higher level of finish is required.

## BUFFING AND POLISHING

To polish your metal to a high luster, you must remove scratches of increasingly finer degrees to smooth the surface and maximize its reflective capabilities. Waves in a piece of metal are a result of high and low spots that should have been removed earlier in the finishing process. Block sanding or filing while the piece is still unfinished reveals high and low spots. If unacceptable waves show up once you start buffing, block sand with a paper just fine enough to scratch the surface for purposes of gauging its flatness. Bump highs and lows into position with a hammer against a clean, blemish-free surface, preferably with some give to it, such as wood.

There are several factors to consider that affect the speed and quality of your buffing and polishing: the aggressiveness of your buffing compound, the type and texture of the buffing wheel, the speed of your wheel, and the manner with which you hold the piece against the buffing wheel.

We use products from Caswell Plating (www.caswellplating. com) and generally follow the recommendations found on their website. Another excellent resource is Jeff Lilly's *How to Restore Metal Auto Body Trim*.

Although Caswell differentiates between buffing with a pulling, cutting stroke and a pushing, coloring stroke in reference to whether you are trying to remove scratches or highly polish a piece, I confess that I like to buff back and forth from side to side most of the time, but I will buff lengthwise if a piece is long and narrow, such as a fender spear. I push harder against the buffing wheel if I want to be more aggressive, and I go to a finer abrasive on a softer wheel with lighter pressure if I want to polish more highly. Always clean off your work with lacquer thinner or a good solvent before switching abrasives. If you fail to do this, you will contaminate your buffing wheels and the strongest abrasive

will always dominate. Likewise, keep ferrous and non-ferrous buffing wheels and materials separate to avoid inducing unwanted scratches in soft metals, and dedicate each buffing wheel to a single abrasive.

For ferrous metals, including stainless steel, we usually start our cutting with the sisal wheel—it looks like a stiff bristle brush—with emery abrasive. Then we go to a ventilated wheel with stainless abrasive and finally polish with a loose section wheel with white rouge. For non-ferrous metals, like brass, copper, and aluminum, we use the sisal wheel with emery compound if needed for heavy cutting, then the spiral sewn wheel with tripoli compound, and polish with the loose section wheel with white rouge. For nickel-plated and chrome-plated parts, we start with the spiral sewn wheel and Tripoli, then polish with the loose section wheel and white rouge.

At left is a brick of Tripoli abrasive with a spiral sewn buffing wheel. At right is a brick of white rouge abrasive with a loose section wheel.

At left is a brick of Tripoli abrasive on a ventilated buffing wheel. At right is a brick of stainless compound on a ventilated wheel.

Coarse emery abrasive is shown with a sisal buffing wheel.

Our 8-inch buffing wheels are mounted on Sears Craftsman bench grinders that have had the guards removed. I have never bothered to calculate the buffing speed of these wheels in surface feet per minute as recommended by Caswell, but I can attest to their effectiveness. Some days they are in use for a couple of hours at a time, and countless projects have been successfully completed using this basic setup and following the above recommendations. Since we last purchased equipment, Sears has begun offering 0.75 horsepower 8-inch bench buffers, but if you don't do a lot of buffing it may make more sense to just convert a bench top grinder at those times that you need it.

Buffing can be dangerous, so please take the following precautions before you begin. Wear a face shield, a dust mask, ear plugs, and cotton gloves. Segregate your gloves according to the buffing material with which they are used. Always keep your piece below the midpoint of the wheel while buffing, and keep the wheel rotating off the edge of your piece, rather than into it. Tape up the nut on the buffing wheel shaft to keep it from inadvertently scratching something. If you don't follow these simple precautions, you will be astounded by the force with which one of these wheels can propel something across the room or into your person. Buffing wheels also like to grab long pieces of irreplaceable trim and wrap them into pretzels—it's uncanny, really. For delicate objects, tape them to a piece of wood to secure them while buffing.

To begin buffing, turn on the machine and press the compound against the wheel momentarily. The wheel will be spinning so fast that it will easily coat itself almost instantaneously. Press the piece against the wheel to obtain a good cutting action, but don't press so hard that you hear the buffer's motor slow down. Move the piece back and forth for about 10 seconds, remove it, and check your progress.

Be safe when buffing. Wear face protection, a dust mask, and soft gloves that are segregated by abrasive type. Always keep the work pointing down and below the midpoint of the buffing wheel so that you don't catch an edge.

Continue buffing, adding a little compound about every minute or so. If your piece starts to look black and you are having trouble making progress, you have gotten too much compound on the wheel. Clean off your piece with solvent, and press a wheel rake against the buffing wheel to clean it. In the future you will want to clean each wheel from time to time to keep it soft and free of debris or accumulations of compound.

## ELECTROPLATING

Although nickel and chrome plating are normally thought of as specialized activities carried out by professionals according to strict regulations, there is some limited electroplating individuals can do at home. In addition to their line of buffing materials, Caswell Plating sells kits for various kinds of plating and aluminum anodizing. If you contemplate doing some of your own plating, purchase one of Caswell's plating manuals online. The manuals are very inexpensive and go into detail regarding the materials you will need and the procedure to follow for various processes. I will give an overview of the nickel-plating process my colleague Chris Paulsen followed to give our pedal car headers an extra special finish.

Caswell's immersion electroplating works through the deposition of positively charged metal particles onto a negatively charged work piece. The particles travel from the positive anode to the negative work through a heated plating solution made using distilled, de-ionized, or reverse osmosis filtered water. The flow of electricity that brings about the plating process is provided by a rectifier, one or more batteries, a battery charger, or a combination of a battery and a charger.

At the beginning of each buffing session, clean the wheel with a wheel rake dedicated to the type of abrasive you are using. Clean the wheels occasionally during use as well.

This aluminum teardrop illustrates the transition from 600-grit DA scratches at the rear, to the middle area buffed with Tripoli abrasive on a spiral sewn wheel, followed by the narrow front of the piece which was polished with white rouge on a loose section wheel.

The amount of power needed for plating is determined by the surface area of the object to be plated. Anodes of the plating metal are suspended in a tank of electrolyte alongside the work and wired as described above so that ions will travel from the anodes to the work. As the plating is underway, the water in the tank is agitated to loosen hydrogen bubbles that form during plating that would otherwise interfere with the plating process. For nickel plating, a work piece will need one or more preliminary layers of copper, followed by a layer of nickel. Nickel can be followed by chrome plating to suit individual tastes. We did not follow the nickel with chrome because we like the warmer look of nickel versus chrome.

Although the rusty, pitted bicycle handlebars we started with were very rough in spots, we knew that with enough work we could make them sparkle. To obtain a pristine plated surface, you must start with a flat, smooth foundation. With that requirement in mind, Simon Marriott cut the bars apart, thoroughly bead-blasted the pieces, and reassembled them into a set of tiny headers. Chris Paulsen then went over all

of the surfaces with a soft pad sander using 80-grit paper. The soft pad sander levels out any irregularities and makes existing rust pits less noticeable. Chris kept the pad moving to minimize the tendency of the sander to create flat spots on the curved surfaces. Then he used a DA palm sander and progressed through 150-, 220-, 400-, and 600-grit papers. An additional final sanding phase involved wet sanding with 600-grit paper. After sanding, the pipes' surfaces were a uniform satin finish. Chris took the pipes to the buffing wheel and buffed them with stainless abrasive on a ventilated wheel, followed by additional buffing with a softer spiral sewn wheel with stainless compound. As with an automotive paint job, your plating will only be as good as the preparation. If you take your time in this surface finishing phase, you will not be disappointed by your finished product. Once our headers had a nice gleam in bare metal, we headed for the plating tank.

Chris degreased our headers in a special degreaser recommended by Caswell, rinsed them thoroughly, and prepared them to receive a coat of copper, which acts like a

We leveled and smoothed the headers surfaces with the soft pad sander in the background followed by the DA palm sander.

Our sanding with 600-grit paper left our surface very smooth, so we followed up by buffing with stainless abrasive on a ventilated wheel.

One tank is for copper plating and another is for nickel plating. The 6-volt battery provides the power, and the light bulbs regulate the current to our plating apparatus.

The copper-plated header emerges from the copper coat bath. After a thorough cleaning, it will be ready for nickel plating. See the chapter opening illustration on page 211 for the finished product.

primer for additional coatings. The copper coating improves corrosion resistance and provides a dense foundation for better adhesion of subsequent coatings. If the surface needs additional surface filling, a second coat of copper may be added that can be sanded and polished for a nicer finish.

Because we use 5-gallon pails for plating, occasionally fitting the parts can be problematic. Such was the case with the headers. The solution with the copper plating was to plate one half of each header and then flip it over and plate the other half. With copper, overlapping the plating is permitted. After testing the pH level of the plating solution, the headers were suspended one at a time in the bath, making sure not to touch either of the anodes. Negative leads were connected to the header, whereas the anodes were wired through regulated amperage depending on the surface area of the cathode (header) to be plated. Each header was then plated for one hour on each half. After removal and a thorough rinsing, each

header was wet-sanded with 600-grit paper and then buffed on a spiral sewn wheel with Tripoli compound. The headers were wiped clean with a wax and grease remover and buffed on a loose section wheel with white rouge compound and degreased again in preparation for nickel plating. The sanding and buffing progression is critical for a high-gloss result.

The nickel-plating process was similar to the copper process, except that the plating stages were not allowed to overlap. Chris used two coats of masking lacquer to cover the half of each header that he did not want to plate. He located the seam right next to the weld at the header collector so that it was not noticeable on the finished piece. After plating one half of each header, Chris removed the first application of mask, applied the masking lacquer to the plated half, flipped the part around, and plated the unmasked half. The finished pieces turned out beautifully, and the seam in the plating is almost impossible to see even when you know where to look for it.

# Chapter 10
# Building a Custom Pedal Car

**O**ur pedal car project came about in response to an invitation to participate in a fund-raiser auction celebrating the 80th anniversary of the Ford Deuce roadster at the Petersen Automotive Museum. Various car customizers around the country were invited to modify a standard pedal car from Warehouse 36 to showcase their creativity and skills and to pay tribute to this classic automobile, the '32 Ford roadster. The pedal car project was fun and challenging at the same time. The work involved was very much like one would expect to carry out on a full-size car, but the reduced scale meant much less time and fewer materials were needed to finish the project. One benefit I had not anticipated was how much creative problem solving a project like this demands and how quickly one can work through numerous solutions to each new problem. Fortunately, we learned a lot from our mistakes. Do not hesitate to undertake a free-form project like this. You will be hard-pressed to find another way to learn as much as quickly. Even if you do not have any interest in pedal cars, building a small-scale version of a car body is

a great exercise. We have since built a small-scale Model A roadster pickup truck, but we ignored all of the pedal car mechanics, and it was well worth the effort.

Although the stock pedal car we began with was fancy for a toy, it cried out for modification. The car's ride height was unacceptably tall and the rear of the body was much too square for our tastes. For inspiration, we looked to a famous hot rod that belongs to our school, a Ford roadster that was modified in the late 1940s by a Californian named Paul Harris. In 1950, Harris took the flathead-powered roadster to Bonneville, where he was clocked at 131 miles per hour. Harris later drag raced the car through the 1950s and sold it in 1958. The roadster was raced, shown, and finally rebuilt in 1972 by James Handy, Andy Brizio, and Jack Hageman. As a tribute to the Paul Harris roadster, we incorporated some of the original car's features into our pedal car. The most distinctive elements of the Harris car in its early years were its track nose, custom long exhaust headers, and steel wheels with Firestone tires.

The complete transformation of this pedal car required many of the same skills one needs to restore full-sized cars—paint, trim, sheet metal, and fabrication skills—plus, it was great fun.

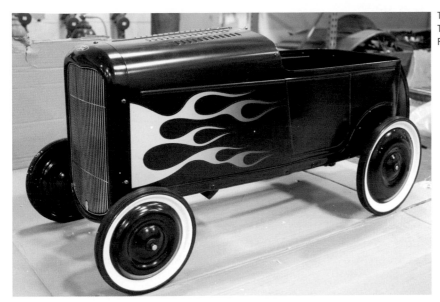

This is the original pedal car that each builder received. The front end of the car looks pretty faithful to the '32 Ford, but the rear of the car is too boxy.

We looked to this unusual hot rod as our inspiration. In the 1950s, it had chrome exhaust headers, steel wheels, and the hand-formed track nose that it still wears today. We decided to incorporate all of the aforementioned features into our car.

## INITIAL PLANS AND WELDS

From the beginning we knew that the track nose would be the most challenging and important part of our pedal car build. Creating the right look was essential to capturing the personality of the original roadster. Taylor Adams made several sketches to determine the ideal dimensions of the track nose and constructed a plywood buck to use as a guide during shaping. Taylor made the nose from five separate pieces of steel. Most of the stretching was done on the planishing hammer with a flat die on top and a crowned die on the bottom. Some shrinking was carried out along the edges of the deepest crown using the thumbnail dies in the Dake hammer. When building a symmetrical piece such as this, take your time making your buck to insure that each side is

the same, and do not stop shaping until all of the individual panels truly fit the buck. You will be tempted to call it good when each panel almost fits tightly against the buck. The sum of several panels being welded together when they are almost right ends up being very wrong, however. The main reason Taylor's track nose turned out beautifully, apart from his considerable skill in shaping, was because he took the time to make the individual panels sit snugly against the buck before he welded them together. Once he was satisfied with the fit of the pieces, Taylor tack welded them together, checked the fit of the nose against the car once more, and then torch-welded everything together. Taylor finished out his welds using the traditional pick-and-file technique described in the chapter on metal finishing.

245

The track nose is the original car's most distinctive feature, so Taylor Adams built a plywood buck and started shaping five steel panels to make up the nose.

Taylor used the thumbnail dies along a few panel edges to abruptly create some shape and then completed the needed stretching with the planishing hammer.

The nose panels were torch welded together. The center panel was left solid so that we could fine-tune the grille opening with the nose installed on the car.

Once the welds on the track nose were smooth, Taylor rolled an offset along its rear edge in the bead roller to allow the nose to slip into the grille shell opening at the front of the car. Because we did not see any reason to remake the pedal car's hood, we decided to flare out the leading edges of the hood at the bottom to match the distinctive bell shape of the track nose. Taylor hammered the hood corners over a shot bag with a plastic mallet to stretch them out and smoothed them by planishing them against a steel dolly with a steel hammer. The finished result looks beautiful and will no doubt cause some people to shake their heads and wonder how this effect was achieved. The lesson to take away from this is the following: sometimes the best way to fix something is to hit it with a hammer. It sounds barbaric, but you must not forget the simple solutions as you become more sophisticated in your shaping.

Simple solutions came into play with the opening in the track nose for the grille as well. Taylor determined the correct grille shape by making several paper cutouts and placing them over the nose. After tracing the best sample, Taylor cut the opening with aviation snips, leaving a half inch of extra metal for a flange around the perimeter of the hole. Most of the flange was bent with a pair of Vise-Grip pliers. Stubborn areas that needed a lot of stretching were heated red hot with a torch and hammered over using a hand-held steel dolly. The flange was cleaned up by gently planishing it with a hammer. Taylor drilled holes around the outer rear edge of the nose and plug welded it to the hood. Quarter-turn Dzus fasteners were added to the sides of the grille shell to imitate those on the original Paul Harris roadster. The safe way to add Dzus fasteners is to rivet them in place before painting your project. You will have a lot more leeway to correct mistakes in the metal working phase than after your panels have a flawless paint job on them. Once all of your panels have been test fitted and you are satisfied with the alignment and number of fasteners, you can drill out the rivets holding the fasteners in place and reinstall them after paint.

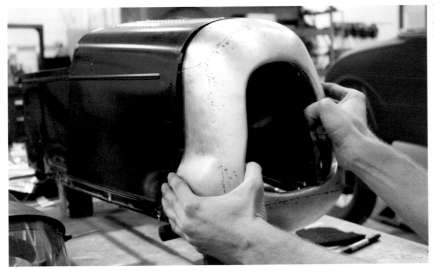

The track nose was run through the bead roller to put an offset bead along its rear edge so it would fit inside the original pedal car grille shell opening.

Taylor created the transition from the bulged track nose to the hood by stretching the hood sides into a shot bag with a hammer. This would work well on a full-sized car as well.

The opening for the grille was cut with aviation snips. A soft inner flange was bent with Vise-Grips by hand. Heat was applied at the stubborn lower corners to facilitate stretching.

With the track nose complete, our pedal car was headed in the right direction, but it looked a little bit too much like an old Korean War–era MiG fighter jet fuselage on the tall stock chassis. Different wheels and tires were selected, the rear axle was moved above the frame, and custom front spindles were fabricated to lower the car several inches. At the front of the car, the frame and steering components were widened 2 inches to allow the body to come down—*channeled*, in hot rod terms—over the frame. As with a full-size car, we found that small modifications often necessitated sweeping changes to retain functionality.

## FABRICATING THE EXHAUST SYSTEM

Lowering the body and identifying the proper running gear established the final ride height for our car. Now we could safely proceed with the fabrication of the exhaust system. When bending the small-diameter tubes proved difficult, we rounded up several sets of old bicycle handlebars, bead-blasted their chrome plating away, and cut them into usable lengths. A 2-inch-diameter length of automotive exhaust pipe was used for the header collector. To affix the headers to the car, bolts were welded to washers having the same diameter as the inside of the individual pipes. The washers were welded inside the pipes, and a bracket was added between the collector and car's frame for support.

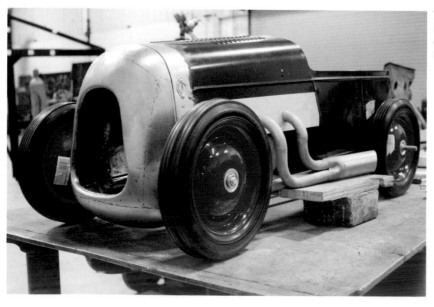

The original pedal car's right height was unacceptably high, so new longer front spindles were milled out of square stock to lower the front of the car. Special lowering blocks were made for the rear axle.

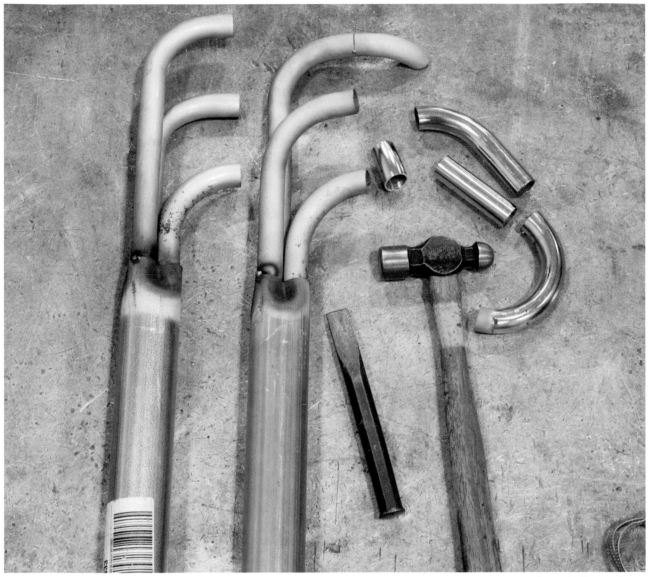

Our headers began life as a couple of sets of rusty bicycle handlebars. Simon Marriott cut the bars into usable curves, bead-blasted the pieces with a couple of lengths of automotive exhaust pipe, and assembled the hodgepodge into a convincing pair of headers. The indentations between the pipes on the collectors were made with a blunt chisel while the metal was red hot.

Our progress on the track nose and exhaust made the rear of the car almost painful to look at; it reminded me of a mailbox. Fortunately, we were working on our pedal car within a few feet of a Model A roadster, so we were able to siphon off inspiration and ideas from the old Ford and channel them into the pedal car. We agreed that the pressed beads suggesting body lines looked pretty convincing on the stock pedal car, but the proportions of the rear of the car were wrong. By taping a piece of template paper over the original pedal car body and rubbing it with a grubby finger, we were able transfer the body lines to our paper as an aid in redesigning the rear quarter panels. Then we traced the outer diameter of the rear tire and the outer contour of

the detachable trunk section onto the same piece of paper. We redrew the door and body lines to harmonize with the new tire size, lower ride height, and lengthened quarter panel. By punching holes in our drawing, we were able to transfer our new ideal dimensions to a steel blank and roll the necessary beads with the bead roller. We allowed for an extra ¼ inch of metal for a flange along the top and bottom edges of our quarter panel for several reasons: first, we needed to be able to generate a gradual curve in the entire panel by shrinking these flanges, we needed to be able to finish off the edges of the panel, we welcomed the additional stiffening the flanges would add, and a flange along the wheel arch area would make it easier to weld in a separate wheel arch.

The bead roller was called upon to suggest door outlines and original-style body lines in our new quarter panels. Short flanges were left along the top of the panels adjacent to the cockpit and in the wheel openings.

Whenever you need to roll several beads in a panel, consider the ideal sequence to follow when rolling before you get carried away. Often there will only be one way to get the intended result without the beads interfering with one another. Also, pay attention to the pressure you apply so that you will be able to replicate your bead depth. On a machine like ours, we simply counted the number of turns on the screw that applies pressure to the upper roller. After we rolled the prominent beads in our quarter panels, we ran the edges through the burring dies on the Pexto beader to create a bend line for the flanges. The flanges were turned the rest of the way with Vise-Grips and by hammering them against a dolly. By shrinking these flanges in the hand-shrinker machine, we created a curve in our quarter panels just like on the body of a full-size Ford.

The stock pedal car trunk bears the Ford script and looks nice, so we decided to incorporate it into our car. The cockpit of the original pedal was exceedingly long, however, so we decided to create a tonneau panel to go above the trunk, resolve the rear of the cockpit, and to provide a transition between the rear quarter panels and the trunk. The bead that

runs along the rear of the passenger compartment of a '32 Ford roadster is complex; it has several bends and curves gradually across its length. When faced with a daunting metal shaping problem like this, try to break it down into smaller digestible parts. After several test samples, we created a reasonable facsimile by splitting the bead up into three parts. The weld seams were chosen based on where we would need a flat edge to shape the metal in the shrinker and stretcher machines and how accessible the seams would be for grinding. The tonneau panel itself was created using the English wheel, and flanges were added along its sides in anticipation of attachment with plug welds.

A subtle but important trick was needed along the front edge of our tonneau panel that is worth mentioning. If you have tried to bend a flange in a curved panel, you have noticed how the bending of the flange tends to pull the shape out of the panel, especially if the flange bends up in opposition to the overall curve on the panel, as here on our tonneau panel. This is because the tension of bending extends out into the entire panel. If you can stretch the flange horizontally before you bend it, however, the tension created in the rest of the panel will be greatly reduced. On our panel, we used the linear stretching die in the planishing hammer to stretch the front flange and then hammered the flange over a curved steel stake in a vise to round its shape. Internalizing this process of stretching and then shaping will help you fabricate increasingly complex panels. This process is exactly the same as the one used to create the reverse curve panel in the advanced shaping chapter.

David Berg put a general crown into the tonneau panel with the English wheel and then put flanges on the sides with the bead roller. By shrinking the flanges with the hand-shrinker, David recovered any of the shape that was pulled out of the panel when the flanges were bent. The flange on the leading edge points up and gets stretched, however.

To join the rear quarter panels to the trunk and tonneau panel, we made some narrow transition pieces about an inch wide. These were basically lengths of metal bent at a 90 degree angle and hand shrunk until they matched the curve of the trunk. Then we rolled a half-round bead in them and welded them to the quarter panels.

To tackle the seam where the rear quarters meet the trunk and tonneau area, we created two long right triangles of metal, bent a ¼-inch flange at 90 degrees along the straight edge, and used the hand shrinker to curve each panel until it matched the curve of the trunk edge. We rolled a bead along the flanged edges so that these panels would resemble those on a '32 Ford and then shrunk the non-flanged edges to curve them down to meet the quarters. Our design placed the weld seam along the apex of the curve running down the rear edge of the quarter panels, which made the weld beads easy to grind. Using a torch-weld on one side and a TIG weld on the other meant that the beads were soft enough to shape and smooth so that the weld seam would be indistinguishable from the surrounding metal.

## FABRICATING THE WHEEL ARCHES

The wheel arches were ideal hammerform projects. We traced the wheel openings in the rear quarters onto four pieces of MDF and hammered out two steel blanks, allowing for a ¼-inch flange around their perimeters for attachment. By

plug welding the wheel arches to the quarters, we obtained a clean-looking finish without having to do a time-consuming butt weld. Plug welds were also used to join the quarters to the car and to the tonneau. The raw edge around the perimeter of the cockpit area was finished by welding a ¼-inch-diameter steel rod along the entire edge and grinding the welds smooth. The end result looked very clean and was much less work than if we had tried to wrap a wire inside the sheet metal along the edge.

The grille was formed from several small strips of aluminum. First the outer perimeter was bent to shape and welded. Then the interior bars were shrunk by hand along their rear edge until they curved, and then they were welded to the rear edge of the grille's perimeter where the weld would not be seen. Helping a frustrated student with this process reminded me of why aluminum has a reputation for being difficult to weld. In this case, fitting each vertical strip of aluminum was a huge hassle. As a result, the fitment where the ends of the aluminum strips met the perimeter of the grille was not bad, but it wasn't perfect either. With

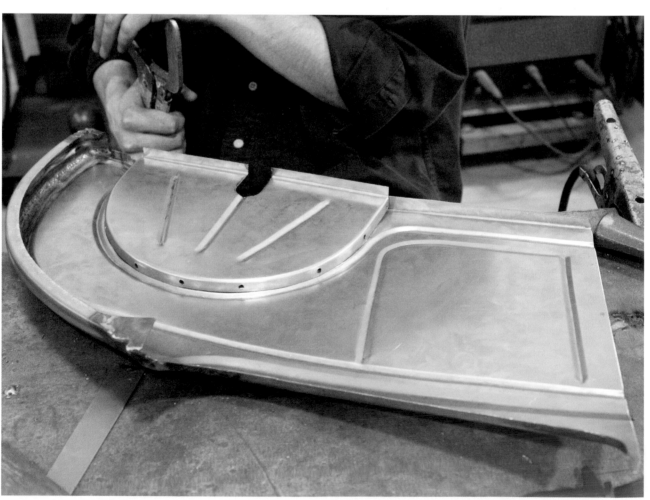

The wheel arches reaffirm the clean machine-made look that hammerforms can produce. We made the basic arches and added beads with the bead roller afterward.

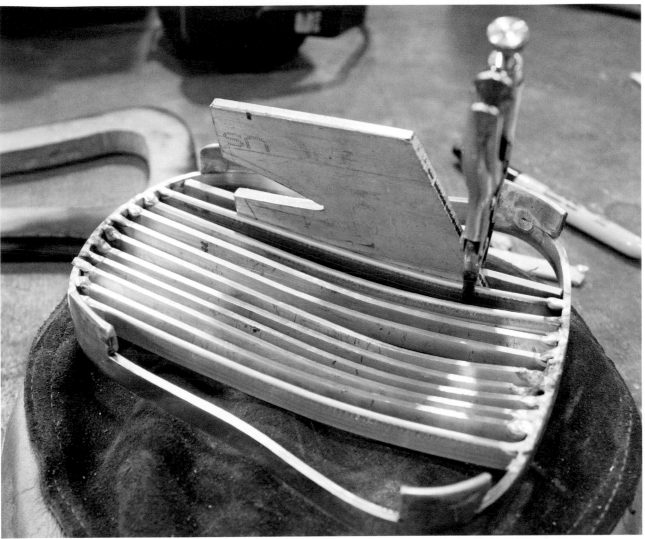

Clamping a scrap of aluminum between the grille bars during welding insured that our spacing was equal between them, and it helped lessen distortion.

aluminum, whenever there is a gap or slight misalignment between pieces to be welded, the TIG arc will always start and stubbornly cling to the first surface that it comes across. For unsuspecting victims hoping to weld successfully, the tendency is to hold the torch on the first puddle that forms and hope in vain that a puddle will form on the second piece of aluminum as well so that the two can be welded together. The first piece becomes overheated and starts to sag without ever fusing to the second piece. I suffered quite a bit to learn the following lesson: in the case of a less than perfect fit, add a little filler rod to whichever piece will puddle. Stop the arc for a moment to let the puddle solidify, then pour on the heat and point it at the piece on which you are having trouble forming a puddle. Ironically, the solution to most aluminum welding problems is to add more heat by stepping further on the pedal or you have a shielding gas issue. If you are having trouble getting aluminum to weld because of a gummy oxide

membrane that forms on the weld pool, perhaps the shielding gas is not set properly, the metal is dirty, or you are letting an oxide-covered ball form on the end of the filler rod before dabbing it into the weld pool. Try to keep the filler rod near the TIG torch for the benefit of the shielding gas, but don't let the rod heat up enough to form a ball on its end.

The idea for the steering wheel for our pedal car was a stroke of genius. Glenn Herman sandwiched a flat plate of aluminum between two large steel cylinders approximately 8 inches in diameter and placed a smaller solid cylinder in the exact center of the plate. By applying pressure to the center cylinder in the hydraulic press while holding the outer perimeter stationary, Glenn pressed out a perfect blank for a steering wheel. This technique is similar to the one we used on the endcaps of the small gas tank in an earlier chapter. He then drilled the spokes, milled a centerpiece, and crafted walnut grips to go around the wheel's rim.

Using the hydraulic press technique that we demonstrated in the gas tank chapter earlier, Glenn formed the steering wheel center section.

The finished steering wheel will no doubt cause people to wonder how it was made.

The pedal car is just about ready for paint.

# Chapter 11
# Floorpans, Rocker Panels, and Rear Quarter Panels

From this point forward in this book, I would like to address major areas of automotive sheet metal repair and look at ways of resolving common problems that arise. First I will discuss floors, rockers, and quarter panels, as those are almost universal problems. I will examine these problems in their order of increasing difficulty.

## FLOORPANS

Floorpans are easy to repair and will provide a surge of satisfaction and a boost to your self-confidence once completed. Furthermore, finished floorpans are tangible proof of progress on your project in case other members of your household covet your garage space. Before you begin, honestly assess the extent of the damage to your floors and the adjacent sills or inner rockers. If the damage you face can be repaired by tracing around a small patch panel and butt welding in a replacement, make sure there are no critical lines or wires nearby and forge ahead. If, however, the support structure along the outer edges of the floor is going to need to be repaired, you might be better off tackling the floors after you have fixed the main structural members along the sills. You do not want to be in the position of installing a new floorpan that cannot be attached along its length because the sill is bad.

For small floor repairs, decide if you want to butt weld a patch in place that will be indistinguishable from the factory appearance or if you are willing to flange one piece over its neighbor after you've cut out the damage. Overlapping one piece of metal over another by flanging it along its length cuts down significantly on fit-up time and may or may not stray from factory appearance, but it increases the chance of corrosion because water can get inside the flange.

The relationship of the panels on the right side of this Ford Mustang look really nice, but it wasn't always this way. The clean lines and even panel gaps are artificial constructions created by the restorer looking for tasteful over-restoration.

The Ford Mustang floor design is typical of many cars of its era. Ideally, you will want weld-thru primer between any overlaps, a good coat of primer on top of the exterior surfaces, and a layer of seam sealer over any panel joints.

Nevertheless, plenty of cars were designed with this feature; an overlap does add stiffness, after all. Overlapping new metal over corroded old metal is just dumb because the overlap will accelerate the corrosion already underway. If your floorpan or patch panel has an overlap around its perimeter, make sure that all surfaces are rust free, that they are coated with weld-thru primer before welding, and that they get coated with seam sealer before the final painting.

For large floorpan repairs, make sure the car is well supported so that nothing can shift out of position, check underneath the car for anything you don't want to cut by mistake and consider removing as much of the interior as possible. Cutting, welding, and grinding makes a huge mess and poses a threat to glass and soft trim. Taping a large piece of cardboard over the windshield and rear glass is a good idea. I will admit to taking the extra precaution of draping a welding blanket loosely overhead in particularly delicate situations. Try to avoid breathing welding fumes. Drill out any spot welds securing the original floor and remove it with an air chisel. Do not throw anything away until you have installed the new floor. You may need to refer to something on the old floor later. Repair and thoroughly clean any surfaces beneath the original floor that need attention, and coat everything with weld-thru primer.

As you fit the new floorpan, check its alignment with the shifter hole, brake master cylinder opening, or clutch slave cylinder opening when these components are present. If the pan will be plug welded around its edges, punch the needed holes and apply weld-thru primer. Compare the new floor with the old to check for seat and seat belt mounts, shock absorber mounts, threaded fittings for exhaust hangers, and hole placement generally. When you are satisfied with the fit of the new floor, secure it with a few self-tapping sheet metal screws so that it will not move when you begin welding. I believe the easiest way to weld in a large floorpan is to treat it like a large piece of cloth that I am trying to stretch tight. Start with welds in the center or center of the longest sides, and work your way out to the corners to chase the slack out to the edges. To obtain a clean, tight, weatherproof result, you must not have any gaps between the floor and the structure to which it is welded. Do whatever is necessary to push the pan down tight to the supports beneath: screw it down, wedge a piece of wood between the dash and floor, or lean over and press the floor down with a hammer wedged against your shoulder as you weld. Grind any welds that stand proud, reweld any pinholes that show themselves, and vacuum out the left over grit with a shop vac. Depending on what repairs lie ahead, you might at least spray the raw metal with some rattle can self-etching primer to prevent rust. If there are no other repairs to make, apply seam sealer to all of the joints and paint the floor. When aftermarket pans are available, floorpan replacement does not involve elaborate fabrication, just patience in fitting, so there is no reason to be reluctant to undertake this project regardless of one's level of experience.

## ROCKER PANELS

Rocker panel repairs are close cousins of floorpan repairs—a repair to one often goes hand in hand with a repair to the other—but they can be slightly more involved because the end product is so open to scrutiny. Typically, rockers are rusted out, but occasionally they are crushed by poorly placed jacks, mauled when a car rolls part way off a trailer, or hit in a collision. Spot repairs to rocker panels are complicated by the difficulty of getting behind a weld seam on a rocker to stretch out low areas caused by shrinkage during welding. As an example of a small rocker repair, I have illustrated a patched Ford Mustang rocker panel. As we anticipated, the metal was slightly low on either side of the MIG weld bead that runs vertically across the rocker. An adjacent long horizontal weld was perfectly flat, however, because the metal was very stiff near the top edge of the rocker. To fix the low area next to the bead, we first attached studs with a stud gun directly on the bead and tried to pull the area along the bead up. We failed to obtain much movement before the studs snapped off. After welding a nut onto the bead, however, we were able to pull the lowest spot to within $\frac{1}{32}$ inch of flat, which was good enough to proceed with the grinding of the weld bead.

In contrast to the Mustang floor, this MGB GT floor is made up of two separate sides and a tunnel. All of the panels get plug welded together in imitation of factory spot welds. Do one side at a time if possible.

Attempting to patch a rocker will usually result in a low area along the weld bead resulting from shrinkage no matter how carefully you set up you panel. The metal will be very rigid too. We made some good headway by welding a nut on the bead at the lowest point and pulling. Attaching the nut to the bead keeps you from marring the pristine sheet metal.

Corrosion damage to rocker panels can be both deceiving and disheartening because you have no way of knowing how extensive the damage is until you have removed the outer rocker. Whenever I have deemed it necessary to remove a rocker panel, I have usually found the ugliness I was expecting, but not always. If there are only a few rust bubbles on the outermost rocker, you may get by with only replacing that panel. If the outer rocker has massive rust holes, however, you can count on needing to replace the inner rocker as well. More rigorous structural support for your car will be needed during the repair of both inner and outer rockers, so plan to stabilize the A- and B-pillars somehow. If the interior of your car is nice and you don't want to weld anything across the door opening, determine if there is a sneaky way you can remove the interior panels and bolt a support across the opening. If you choose the bolt-in method, use two fasteners at each end of the support so that neither end can become a pivot point, and back up the attachment points with large washers to spread the load. Use lock washers on the bolts as well. You may be tempted to bolt some tubing from one or both door hinge mounting points to the door striker mounting point, but I do not recommend doing this because you must be able to check the door fit before you weld the rocker in place. Whenever conducting one of two similar repairs, such as a rocker or quarter panel, always repair *one side at a time*. Leaving one side undisturbed maintains structural integrity and serves as a valuable reference for seeing how everything fits together.

With your car's structure adequately reinforced and your door fitting the best that it possibly can, begin the rocker repair procedure by scrubbing the perimeter flange with a piece of sandpaper by hand. Your fingers will allow the sandpaper to follow the undulations of the flange, making the spot welds retaining the outer rocker easy to see. Drill out the spot welds with a spot weld bit and use an air chisel to separate the outer rocker from the inner rocker with as little drama as possible. You will know right away if further repairs are needed. Depending on the model of car, the outer rocker may be the only panel that needs replacement. If the inner rocker is satisfactory, clean it thoroughly with a wire brush chucked up in a drill and coat it with weld-thru primer.

Remarkably, this MG Midget inner rocker can be saved with some diligent cleaning and some repairs above the rocker. Often the inner rocker is as bad as the outer. Note the support across the door opening.

If the inner rocker is a mess, double-check that you have added enough reinforcement to allow for its removal and drill out the spot welds holding it in place. There is nothing especially difficult about removing the inner rocker. Just make sure brake lines, fuel lines, or wiring are not in the way. Grind off any primer from areas that will be welded, and spray them with weld-thru primer. Secure the inner rocker in place with self-tapping sheet metal screws or tack weld it in place temporarily.

Drill or punch ³⁄₁₆- to ⁵⁄₁₆-inch-diameter holes around the perimeter of the new outer rocker panel with the same spacing that was used on the original rocker and grind off any of the black electrostatic primer that may exist on the flange. A pneumatic punch like the one Eastwood sells is well worth its cost in an application like this. Spray the back side of the flange with weld-thru primer and clamp the outer rocker in place with Vise-Grips situated in such a way that you can still close the door. Now check the door fit. Sometimes, aftermarket rocker panels are pressed using inferior dies, with the result that the new panels lack the definition of the originals. Do not be surprised if the first bend where the rocker meets the door bottom is too soft and makes the door gap appear excessive. I remember running into this problem with Austin-Healey reproduction rocker panels. Adjust the fit the best you can and screw the new outer rocker in place along its top edge with self-tapping sheet metal screws. With the door closed, ask yourself how the gap looks. If the door gap is too wide, place a long piece of wood along the bottom of the rocker and push on it with a floor jack. This trick will usually make up the last bit of difference between looking fair and looking good. Run some self-tapping sheet metal screws through the bottom flange of the outer rocker to hold it in place.

If you placed a jack under the rocker, remove it and step back to admire your work from a distance. If you hear angels singing or receive some other internal acknowledgement of excellence, begin plug welding the rocker in place with a MIG welder. If you do not have much experience with plug welds and weld-thru primer, try a few practice welds. Weld-thru primer may seem at first like weld-proof primer. The weld will spatter quite a bit, and the arc will naturally gravitate toward the upper layer of steel, rather than the bottom layer where you want it. Do not let the weld pile up in the hole. Instead, create a weld pool on the lower layer that you can swirl around the perimeter of the hole to the point that the top layer fuses to the bottom layer. When you let off the trigger, the weld should be flat to minimize grinding and there should be no voids or pinholes in the weld. If the weld-thru primer gives you fits, chuck up a wooden dowel the same size as the hole in an electric drill and briefly spin it in each hole to scrape off the primer. Welding the clean metal is infinitely easier than welding on the primer. Alternate your welds at different points to avoid pouring tons of heat into the panel and to keep it from pulling itself out of alignment.

## QUARTER PANEL REPAIRS

As you spend some time getting to know your outer and perhaps inner rocker panels, you may notice some corrosion in a rear wheel arch, wheelhouse, or perhaps along a rear wheel opening nearby. Damaged or rusty rear quarter panels and wheel arches are common problems that the car enthusiast must solve. Body designs vary widely, of course, but I'd like to offer some general tips to help you successfully repair your car. Depending on the extent of the damage and the cost of aftermarket quarter panels, when available, replacing an entire quarter panel can be an appealing option. On the ubiquitous Ford Mustang, for example, a quarter panel skin costs about as much as a nice dinner for two, and the fit will be almost satisfactory. With a little adjustment to the flanges at either end of the panel, it will be good. The panels are so inexpensive that you could order two quarters in case you make a mistake. In a moment I will demonstrate how I would go about repairing a quarter panel from scratch, but first I will discuss repairs using ready-made panels.

If you find yourself needing to replace or repair a rear quarter panel, look over the design of the panels involved and determine the scope of the repair—is it a patch or will you be cutting out the entire quarter panel? Ironically, replacing the entire quarter may be more total work, but an easier repair. A quarter panel that reaches all the way to a factory seam obviates the need for welding across a flat expanse of sheet metal altogether, but it may extend up into an area that will necessitate removing the rear glass. The welding on such a repair will not be extensive. There will be plug welds along every edge, but you will not be welding any long butt joints. If the panel fits well, such a repair is not difficult. Unfortunately, accurately judging the fit of a panel like this is just about impossible without actually putting it in place on the car. Therefore, just as you would do with other structural repairs, support the body so that nothing will shift out of alignment once the old panel is removed. Melt away the factory lead as needed at the body seams, drill out the old spot welds, pry the old panel loose with an air chisel, and remove it without using any more force than necessary. Try to determine if the new panel will work satisfactorily by holding it in position. If you tamper with it you won't be able to return it, of course, so make an educated guess at this point if you want to move forward with the new panel. Obviously, the prospect of trying to reuse the old panel is never very appealing, but if the new panel does not fit, fixing the original may be the only option, or try to get one from a donor car.

If the new panel fits, grind away the black electrostatic primer at the seams, punch holes for plug welds, apply weld-thru primer to the joining surfaces, and use Clecos or sheet metal screws to secure the panel for welding. After welding, grind the welds as needed and apply lead to the seams at the C-pillar and in the tonneau area and the panel will be just like it came from the factory. If you will be removing the glass from your car anyway, removing the full quarter panel might

The design of this 1954 Plymouth Belvedere wheelhouse is common to many cars—a two-piece wheelhouse/wheel well that is resistance welded down the center and at its edges.

This MGB GT quarter panel is typical of cars built from the 1950s through the 1970s. Like the Belvedere in the previous image, this car has a two-piece wheelhouse that can give you access to the back of the weld seam if you need it. Also, placing a long weld seam near a rigid body line will cut down on heat distortion from welding.

be the way to go. If entire panels are not available for your car, or if you don't want to replace any more metal than you have to, you will have no choice but to cut out the cancerous portion of the original and weld in a patch. Some of us gravitate toward the more challenging path anyway, whether it's out of stupidity, insanity, or simply the desire to keep as much of the original car as possible.

If you will be repairing or replacing less than the full quarter panel, try to formulate a logical sequence for the repair. The critical concerns are that the panel fits well against the door and rocker, that it has the right shape, that the wheel opening looks correct and fits the wheelwell/ wheelhouse behind, and that the weld seam can be finished

out satisfactorily. One of the more frustrating aspects of automotive restoration is the need to remove or assemble things in a sequence that the factory never intended—sort of like having to remove the engine in your wife's minivan to get to a spark plug. A painful truth is that at times things were put together in the most expedient manner without regard for their ever needing to be removed, though this is less the case with old cars than newer ones. With that thought in mind, what sequence of events will allow you to put all the pieces back and still satisfy the main concerns? In a perfect world, one would like to be able to proceed as follows: 1) Scribe a cut line on the old quarter panel derived from the edges of the new panel; 2) cut out the old quarter panel, leaving the original wheelhouse in place; 3) tack weld or screw the new quarter panel in position using the original wheelhouse as an aid to placement; 4) cut out the old wheelhouse; 5) weld in the new quarter panel and finish out the weld seam; 6) weld in the new wheelhouse and grind down the welds associated with it; and 7) celebrate the unqualified success of the repair. This sounds easy enough, but the factory attachment methods might require mixing up the order somewhat. If the wheelhouse is made of two halves that were originally resistance welded along a vertical flange behind the quarter panel, for instance, try to get away with adding the outermost wheelwell piece or pieces after you have finished out the long weld seam on the quarter panel. If a two-piece wheelhouse can be screwed together with screws, inserted to check fitment, and then removed at least in part while the quarter panel is being welded and finished, life cannot get any better. You may have to do your welding along the top of the wheelhouse by reaching behind through the trunk or through a hole on the interior side of the car, but a perfectly butt-welded and finished quarter panel is a worthy tradeoff for the minor inconvenience of welding the wheelhouse from an awkward position.

If your wheelhouse is fine and will not be replaced, then you will have the challenge of finishing out the weld seam on the quarter panel with very limited access to the back side of the panel. Once I cut out part of the wheelwell solely to gain access to the back of a quarter panel that was poorly welded by someone else. When the weld seam along the quarter panel was nicely finished, I welded the wheelhouse back together. The bottom line of all of this strategizing is to facilitate the welding and finishing of the quarter panel repair. Regardless of the sequence of steps you decide to take, the actual process of fitting and welding will not change much.

Lay the new quarter panel over the old panel if possible to scribe a trim line along the top edge. Scribe a few vertical registration marks as well to ensure that the new panel goes in exactly in the right spot front to back. A former student of mine, Alex Kaminskas, showed me a clever trick for keeping a long panel such as this in perfect alignment during welding. As he trimmed a new quarter panel skin he was preparing to install, he left a few short tabs along its length. He allowed the tabs on the new panel to overlap the edge of the original panel on the car. He then drilled the tabs and put a Cleco through both panels to hold them in close to perfect alignment. As he approached to within a few inches of the tabs during the tack weld phase, he cut the tabs off with a saber saw and pushed the seam flat for welding. This might be a good option for maintaining perfect alignment for those extra-long panels that are hard to reach with Vise-Grips.

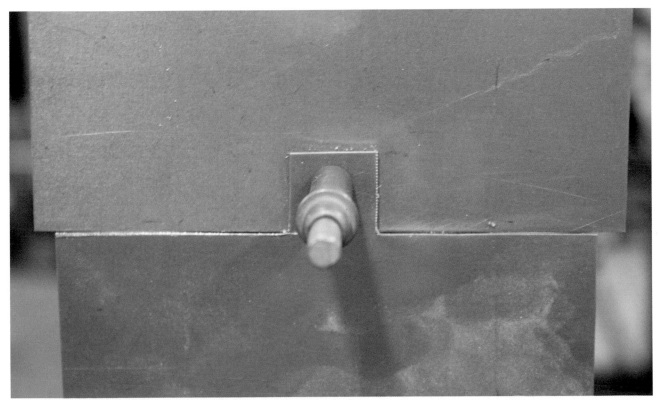
Here is a good way to secure two panels that are hard to clamp. Leave some tabs on one panel when you trim it and stick a Cleco through both panels. After you've established some tack welds, cut out the tabs with a thin saber saw blade.

If I have excellent access to the back of the weld seam, I TIG-weld the panel, but if I have limited access, I MIG-weld it. The TIG-weld is soft and easy to finish, but you have to be able to hammer the seam to counteract the significant shrinking that takes place during welding. Without hammering, the TIG-weld will end up very low. When MIG-welding a new quarter panel, you will definitely want to move around quite a bit and use the blow gun as described in the welding chapter to minimize the heat put into the panel. Do whatever is necessary to keep the edges of adjacent panels in the same plane during welding.

Some people put an offset flange on either the edge of the new quarter panel or the edge of the area to be repaired before welding. They then punch holes in the uppermost panel and plug weld it place. I have never repaired a quarter panel this way, but I would imagine that it takes about a quarter of the time than the method I recommend does. There is nothing illegal or immoral about repairing a quarter panel this way as long as the rear of the overlapped joint gets thoroughly seam sealed, but it seems to me that a thick layer of plastic filler must

necessarily be added to cover the overlapped seam. Plastic filler is not inherently evil, of course, but we all know that the thicker it becomes the less likely it is to stay in place on the car.

If you need to patch a rear quarter panel for a car for which no aftermarket panels are made, do not lose heart. Train yourself to stop thinking of sheet metal panels as standalone entities like "a quarter panel for a Mustang" or a "cab corner for my 1937 Studebaker truck." Instead, think of them as collections of shapes. With a little meditation and green tea, I am convinced that you, too, will be able to break these intimidating, complex forms down into smaller pieces. One of my favorite exercises in class is to show beginning students the corroded bottom corner of a '54 Plymouth Belvedere rear quarter panel. When I ask them how to repair it, I usually receive some puzzled looks and some really involved solutions. When I finally elicit the easiest solution I can come up with—to divide it into two pieces—mental light bulbs click on all over the room. The lesson to take away from this example, therefore, is to try to break your foreign, complex shape into a few simpler ones.

This seemingly complex repair is really not so bad. The short portion of the panel below the blue tape line is a 90 degree angle that has been shrunk and stretched by hand. The rest of the panel is stretched or shrunk as indicated, and another flange has been hammered over along the bottom edge. Break complex shapes into smaller simpler ones.

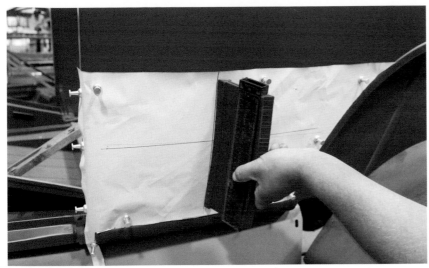

This Ford Model A quarter repair illustrates the basic procedure for patching a crowned panel that has a bead in it. Stick some paper on the panel with magnets, outline the bead with your fingernail, and record the panel's contour with some templates.

Using a very low crown anvil wheel, we quickly establish the correct contour in our patch by wheeling mostly vertically across the panel, augmented with a little wheeling horizontally.

## FORD MODEL A REAR QUARTER PANEL REPAIR

Although I suspect that quarter panel patches are available for the Ford Model A, this panel is good for demonstration purposes because it is a compound curve and it has a bead running along the bottom edge. To replicate this panel, therefore, we will need to copy the curve of the panel vertically and horizontally, the exterior dimensions, and the profile as well as placement of the bead. To obtain a road map of the bead's location, we laid a piece of newsprint over the lower portion of the quarter panel and secured it with magnets. We defined the edge of the bead against the paper with the edge of a fingernail followed by a ballpoint pen, and then we cut off the extra paper outside the line. With the paper pattern still secured to the panel, we placed a contour gauge vertically and horizontally across the center of the curved panel and traced each of those two arcs onto stiff cardboard to use as templates. We marked the locations of the templates on the paper and in relation to one another to eliminate any ambiguity regarding their placement. On this particular panel their placement isn't that critical because the curve of the panel is pretty uniform vertically and horizontally. The curve one way is more drastic than the other, but each curve is consistent for a large area. Other panels may not be so consistent, so it is a good idea to mark your templates' locations in relation to one another and to the panel on which they sit.

Armed with a paper pattern and two templates, we cut out a metal blank and started wheeling our panel with a fairly flat anvil wheel. Most of the wheeling was done in a vertical direction, but we also wheeled horizontally to induce a compound curve and we made a couple of passes diagonally in two directions to even out the panel. When our templates fit the panel horizontally and vertically, we traced the paper pattern on the back of our blank and moved to the bead roller.

Experiment to find the right dies to make your bead and roll it in two or three passes, gradually increasing the pressure from one pass to the next.

To square off the end of the bead abruptly, we sandwiched the target area between two pieces of thicker steel and defined the new edge with a punch made from a damaged rivet set.

When we began this project we had two different Model As in our lab area, and each one had beads with different profiles. We had an easier time matching the bead illustrated here, so that is the one we used for the demonstration. Do not be surprised to find other Model As with slightly different beads. We ran quite a few scraps of metal through the bead roller to find just the right spacing between the rollers and to get an idea of how much pressure to use. We rolled our bead by eye following a permanent marker line, but you can set up a fence by clamping a piece of angle iron to the bead roller and carefully trimming your blank so that the edge can follow the fence at exactly the distance needed to give you the width of the bead plus the flange next to the bead. The front edge of the bead stops abruptly before it meets the B-pillar on the Model A roadster. To square off the front edge of the bead, we sandwiched the sheet metal between two thicker pieces of steel and clamped the assembly to a workbench. The square piece of scrap steel we placed on the underside of the bead served as a male die, therefore, and we were able to define the new front edge of the bead with a punch. With the bead formed, we laid our panel on top of the original and marked the front and bottom edge to locate the bend line for the flange on those two sides.

At the lower front edge of the quarter patch where the wheel arch begins, we determined that a small section of metal needed to be removed to enable us to bend the metal at the corner cleanly. Otherwise, a sharp protruding corner might have formed there. We drilled a small hole there, but we could have made a single cut with a cut-off wheel. When in doubt, fold a piece of paper cut to the shape of the metal blank at the corner to figure out how the corner needs to be modified to be bent successfully. We created a bend line for our flanges with a quick pass through the bead roller. We finished bending the flanges by sandwiching the panel

between a table and a dolly and hammering the flanges over right on the line. We wanted our panel to be able to sit on top of the original panel, so all of our flanges are bent just barely wider than the original. Were we trying to match it perfectly, we would have bent the flanges inside the lines, but that makes for a less successful demonstration piece because we couldn't superimpose the new panel over the old. Remember that when you bend a flange along the edge of a curved panel you often will need to shrink or stretch the flange by hand to maintain the panel's curve. We test fit the panel by placing it on top of the original quarter and considered it finished.

Where the bottom of the panel meets the curve of the rear fender opening, we drilled a small hole to allow the respective flanges to be folded without creating a sharp corner. You could also make a cut the width of a cut-off wheel.

We used a thin upper beading roller against a flat bottom roller to create a bend line for the short flange that runs along the lower edges of the patch. We hammered the flange the rest of the way. Remember that bending a flange on a curve will usually require stretching or shrinking to maintain the panel's shape.

We bent the forward flange in the English wheel by pushing up against the upper wheel.

Because the forward flange curves, we needed to hand-shrink to account for the shorter distance the flange occupies after the bend. Otherwise, the bend will pull the shape out of the panel.

The finished panel looks just right, though to be usable we would have had to bend the flanges slightly inside where they are now. As it stands, our oversized panel fits on top of the original.

## MUSTANG WHEEL ARCH

One last quarter panel feature that I think readers might find interesting is a demonstration of how to make a wheel arch. Whatever your application, whether it be a small fender lip along the outside of a wheel opening, or a full-blown disco-era wheel flare, the techniques required to make it are exactly the same. The first step in creating a wheel arch is to plot out the design. An easy way to do this is to stretch some tape along your intended outline. Tape shows up well from several feet away and is easily changed. For purposes of demonstration, we used a Mustang quarter panel. We found the center of the wheel opening by measuring from the back of the car and marked it. Then we stretched a piece of tape from the top center of the wheel opening to the bottom front of the planned wheel arch. We manipulated the curve until we were happy with it, then placed a piece of newsprint over the tape and secured it with magnets. By scribbling over the tape edge with a pencil and tracing the bottom edge of the wheel opening onto the paper, we obtained a perfect paper outline of one half of our intended wheel arch. We cut this outline out and flipped it over to transfer it onto the rear half of the same wheel opening. We outlined the second half of the arch with tape and looked at it from several feet away to make sure it was what we wanted. This process worked surprisingly well and could be used to duplicate the arch shape on the other side of the car.

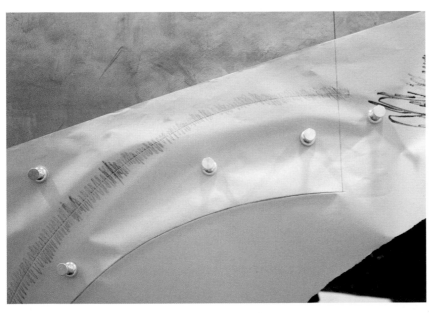

The techniques we present here would apply to any size wheel arch from a stock repair to a modified race car. First, we plot out the shape of one half of the arch on the panel with tape, overlay it with newsprint, and trace the outline.

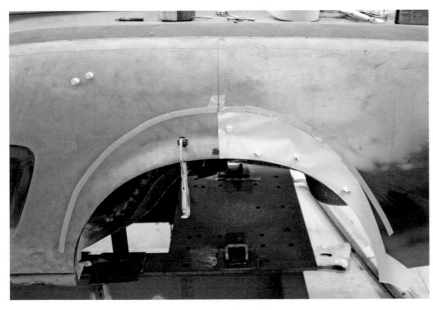

Next we cut out the paper template of the front half of the arch, flip it over and stick it to the rear half of the quarter panel and inspect it. Satisfied with the appearance, we trace the rear template's location with tape. We can use the paper template on the opposite side of the car as well.

There are several possible ways to create the wheel arch shape, either from a flat panel or by enlarging a stock quarter panel, as we did. The bottom line is that a lot of stretching needs to take place along the arch so the metal can occupy the much longer length and larger area of the arch. The fastest way to add length to the metal that will make up the arch is a linear stretching die or linear stretching hammer. The elongated footprint of a linear tool face stretches more to the sides than the ends, which is exactly what we want here. An English wheel would also work very well for this application if you were to wheel in the direction of the arch, but wheeling a panel this large would be awkward, even for two people. If the arch were a separate weld-on piece, however, one person could easily wheel it. If the panel were on the car, one would

have to stretch the arch by hand with a linear stretching hammer struck against a dolly or with an air hammer against a dolly. The air hammer die would most likely be round, rather than linear in shape, but it would still be possible. If you have an English wheel with a narrow 2-inch-wide upper wheel, with an assistant's help you can refine the transition between the arch and the panel by wheeling next the arch's origin as it emerges from the panel. If you do not have a narrow upper wheel, you can use the planishing hammer or do it by hand with a hammer and dolly. Make some cross-templates to match the profile side to side. Use these templates as guides to shape the opposite quarter panel. Once the wheel arch is fully stretched, simply fold the original flange back over with a hammer and dolly.

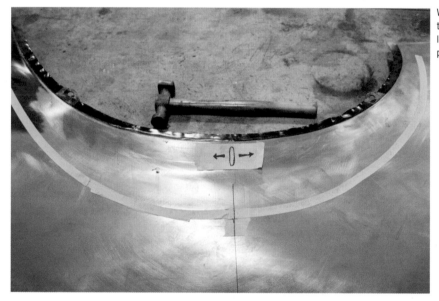

We peel back the short flange that runs along the inside of the wheel arch and stretch the metal over the arch with a linear stretching die in the planishing hammer with the die perpendicular to the curve.

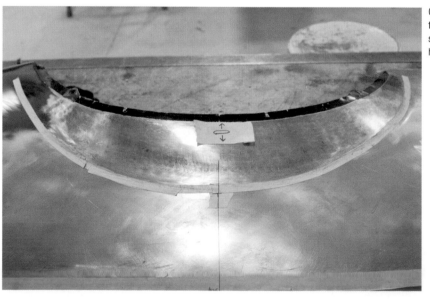

Once we have stretched the arch enough, we rotate the die so that it is parallel to the curve and shape the stretched metal to the desired contour. In conclusion, we hammer the short flange back along its original bend line.

When we are satisfied with the arch's appearance on the car, we could use the profile gauge to makes some cross-templates to use in shaping the panel on the other side of the car.

The enlarged wheel arch is hardly noticeable when viewed independently.

However, when compared to a stock quarter panel, the modification we've just carried out is obvious.

# Chapter 12
# Repairing Doors

## DOOR SKINS

Replacing a door skin is one of the more common sheet metal projects you are likely to encounter if you dabble in the restoration of old cars. For some car models, aftermarket door skins are available, but they vary greatly in quality. If the panels are high quality, re-skinning your door with an aftermarket door skin will be infinitely easier than making your own. Do not be ashamed to take the easier path if it is available. There will be enough challenges in restoring any car, no matter how common it is, to allow you to accumulate plenty of sweat equity in your project. Join a car club or online forum and find out what other people's experiences have been with various suppliers' products so that you can make an informed buying decision. Keep in mind, too, that

one company's panels may originate in several different places depending on the specific panel, so the quality may vary accordingly. When I worked in a shop restoring British cars, I learned which panels were good and which were not from any given supplier, so pay close attention to exactly what products other car enthusiasts recommend.

If good aftermarket door skins are not available for your car, or if you simply welcome the challenge of making your own, I will describe the best procedure I have found for making door skins. First, assess how well the door really fits on the car. Correct body issues with your original door before you start tampering with the skin. Adjust the latch and shim the hinges if necessary to obtain the best fit with the door you have. The status quo contains a lot of

Besides a nasty karate chop across the belt line and some rust along the bottom, Joe McCullough's Model A door was fine. After a couple of tedious hours pushing inside the door with a dolly and tapping on the outside with a hammer, the belt line dent was repaired. The new skin was basically a crowned panel with a bead rolled along the bottom.

information that will forever be lost once you start repairing things. It is much easier to correct pillar and sill problems with an original door for reference, even if the door is rusty, than trying to do so with a hastily re-skinned door that may or may not be right. With the original door in place, ask yourself whether you have to compensate for poor fitment when you make your new door skin. If so, you had better take several pictures and make detailed notes regarding what needs to happen to correct the fit. Next, ask yourself which sides of the door are supposed to be straight; hold a very long straightedge against them to check. If your door curves up at its ends and has an uneven gap where it meets the rocker panel, you do not want to replicate that error in your new skin. Do not assume that anything is correct without checking or you will surely find that it is not correct after you have assiduously copied it. I am speaking from experience.

Assess the condition of your door's hinges and support structure. If the hinges are sloppy, fix these before tampering with the skin. Most auto parts stores sell universal hinge bushing kits that work well in many applications, but you may have to buy replacement hinges or find better originals. If the door's bottom or inner support structures are seriously corroded, fix these issues now. In the case of a rotten door bottom, for example, leave the outer skin undisturbed as much as possible so that you can shape the repair patch to fit behind the original outer skin. The worst thing you can do is to cut out all of the damage and try to rebuild the door in thin air without any points of reference. After all, the original outer skin, rusty though it may be, is the most reliable indicator you have of the correct shape of your door. Usually you can drill out any welds and cut off any flanges that hold your outer door skin to the inner structure behind without disturbing the contour of the skin or violating its pristine edges.

If you end up cutting out any lower portions of your door, consider bead-blasting inside the door while you have access. You can do this as well when the skin is off, but it is worth mentioning that you should always take advantage of any opportunity to clean and paint normally inaccessible areas with weld-thru primer. In some cases, you might want to weld some temporary support, such as a length or two of square tubing, across the back of your door to keep it from springing out of shape when you cut a large piece out of it. The rigidity of your door's design may not be easy to assess, so if you have any doubts, add some braces to be sure, but make sure you weld them in a way that makes them easy to remove. Also, place them in such a way that they will not prevent you from remounting the door on the car in case you need to check its fit during the repair process.

## CUTTING THE DOOR STEEL

Once you are intimately familiar with the fit and condition of the door you intend to repair, determine the size of door skin you will need and cut some mild steel of the appropriate gauge oversize, leaving room for edges to fold over on the sides and bottom as dictated by the design of your door. Even if your damage is only along the bottom of the door skin, re-skin high enough that you can reach the back side of the weld with a steel dolly in your hand. Doing this will greatly facilitate metal finishing the weld. If you are using an aftermarket skin, grind off the factory-applied black primer where the door skin will be welded on its edges and spray it with weld-thru primer on any surfaces that will overlap other metal. Do not spray weld-thru primer along any edge that will be butt-welded to the original metal, however. An easy preventative measure is to run a length of masking tape along the anticipated weld seam before spraying the weld-thru primer. Pull the tape off after painting the weld-thru primer to reveal a clean edge for welding.

For a handmade door skin, shape your panel so that it will sit happily on your original door without rocking, but do not fold over any edges yet. Most doors have more crown from top to bottom than from side to side, so use a relatively flat anvil wheel in the English wheel and do most of your wheeling in this direction—top to bottom, bottom to top, as you work your way across the door with very close tracks. Wheeling will add crown and stiffen the skin so that it will be less likely to flatten out and lose its shape when you weld it in place. Some doors, such as those on the 1950 Mercury in the demonstration, require a little extra trickery than just rolling a straight-forward crown. On the Mercury door, the crown is fairly even across a 10-inch-wide narrow band that runs horizontally the length of the door. Above and below that line the curve increases. This transition is basically a bend line. To create a curve such as this, draw a line down the door to locate the middle of the transition area, then make another pass through the wheel and pull down slightly when you cross the line on your pulling stroke as you guide the panel back and forth through the wheel. Continue wheeling across the panel, bending down very slightly as you pull the metal toward you. You will obtain an even bend similar to if you had pressed the panel down horizontally over a giant cylinder. If your door curves down precipitously at the corners or otherwise, you can use this same trick to influence the curvature of the door. Just pull down where you need to start the bend. Keep in mind that bending one area down will cause an adjacent area to rise up, however. As you begin wheeling large panels like this, your natural tendency when faced with the need for more shape is to wheel more. But keep in mind what kind of shape it is that you need. Wheeling with pressure stretches, thins, and creates a hump as you trade thickness for surface area. Sometimes that is precisely what you need. In the examples I just described, however, you are wheeling with light pressure; you are bending, not stretching the metal.

By now you will have checked the edges and bottom of the original door with a straightedge. If the door edges are laser straight where they meet the body, good. If they are

With an air sickness bag clutched tightly in my fist, I survey the condition of this 1950 Mercury door. It is really, really bad. Because of its large size and ghastly condition, the Merc door makes the perfect demonstration piece.

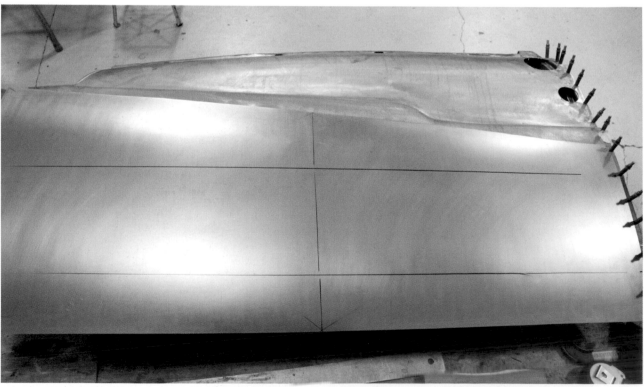

Long panels will stretch themselves out of shape in the English wheel unless they are well supported. Work your way from one end to the middle, then remove the panel, flip it around, and do the same thing from the other end. The center area between the horizontal lines on this door is an even bend from top to bottom. Beyond the lines the bend is more severe.

curved, determine if the curve is supposed to be there. If it is, trace the curve. If a curve doesn't belong, such as along the bottom of the door where it meets the sill or rocker, correct for the curve when you make your skin so that your door gap will be even along the bottom. Take pictures to document shape or weld locations if needed.

With your new door skin superimposed over the old one, draw or scribe a line alongside the top of your skin across the door to identify the cut line. Include a few vertical witness marks to center the new panel over the old so you'll know the exact location when you weld in the panel. Clamp the new skin in place with Vise-Grips, but put something under the jaws so that they don't dent the metal. Scribe or draw a line around the outer edge of the original door to establish where the new edge must be on your handmade panel. This process is a little tricky because a small error will be very noticeable if the edge of the door is not folded over in the right place. At times I have traced the support structure to which the door skin attaches, but I do not recommend that method because the support structure is sometimes quite a

bit undersized compared to the original outer edge of the door. As a result, your door gaps will be too large. Instead, I definitely recommend tracing the original door's contour, but remember to fold the flange inside that line or the door will be too large. When you feel your new skin is accurately marked and a cut line has been transferred to the door, turn the door face-down and grind the edges of the old skin until they turn blue with a hand-held 5-inch grinder. Don't grind all the way through the metal. Preserve as much of the original inner door structure as you can. By grinding the extreme edge whisper thin, you will have no trouble peeling away the flange that runs around the perimeter of the skin. Use a flat-bladed screwdriver or scraper bent at a 90 degree angle to separate the lip of the door skin where it folds over the back of the door gently from the inner door structure. If possible, repair or replace any of the inner support structure without disturbing your outer skin—remember: this is your guide for welding new metal back in. If your repaired inner door structure fits the original outer skin, your door is correct thus far.

The old door skin can be easily removed by grinding the edges of the skin until they are very thin and peeling back the metal with a homemade prying tool.

Don't grind away too much metal or you will damage the support structure underneath the skin. With the old skin separated, repair the understructure if needed, clean it, and coat it with weld-thru primer.

## CUT THE DOOR

Decide what type of welding process you will use to weld the new skin on and cut across the door. Cut next to the trim line for a TIG or torch-weld, and cut on the trim line for a MIG weld—you will want about a 0.040-inch gap for the MIG. Try to make your cut as straight as possible. Cutting along a piece of masking tape is sometimes helpful for cutting a straight line. Any sideways movement of your cut-off wheel will create a jagged edge that will be a hassle to even up. Remember to bead-blast, grind, wire brush, or in some way clean the inner door structure and coat it with weld-thru primer.

Position your new skin on the door. Setup is at least 80 percent of the battle. Carefully align your skin in place. If the top edge needs adjustment, a large file pulled along the edge with a gloved hand works well. Normally files are used with a pushing stroke, but if you turn the file around backward and pull it toward you along the edge of the trimmed original door skin, you will get a magnificently straight, clean edge. Wear an oven-mitt style welding glove, however, because the edge is sharp. When the perfect fit of your new door skin takes your breath away, glance at the edges of the support structure on the back of the door skin to make sure there is enough metal around the perimeter of the new skin for the flanges to bend over. Consult your notes and pictures for guidance on any changes that might need to be made to the edges of door. Once you are satisfied that you know where the edges should be bent, cut the bottom corners of the skin at 45 degrees for a mitred, picture-frame appearance once the edges get folded over. I always cut a piece of paper in the exact shape of each lower door corner and fold its edges to find out exactly where to trim the skin. Then I trace the paper template and trim accordingly. Now you are ready to put flanges around the perimeter of the door skin.

There are a few ways to bend the flange into the door skin's edges. You can make several passes through the burring dies on the bead roller, lifting the panel up a little farther each time. You can bend the flange on the English wheel. You can bend the flange a little at a time with Vise-Grip pliers, or you can squeeze the skin between a dolly and the table or between a spoon and a table and hammer the flange over. The looser the clamping, the more the act of bending the flange will pull some crown out of the skin and soften the bent edge—neither of which is desirable. Any time you attempt to bend a flange along the edge of a crowned piece of sheet metal you will be asking the length of metal that will make up the flange to fit into a shorter space. This is a critical concept that is worth restating. Because the flange on the edge of a crowned door panel must curve to maintain the panel's shape, shrinking must take place along the flange or the act of bending the flange will pull the shape out of panel. The metal does not want to be forced into the shorter curved arc the flange represents. Once the flange is at 90 degrees to the skin, you can restore the curvature in the skin by shrinking the flange with a hand shrinker, or alternate shrinking with bending as you create the flange in the first place. Drill or punch holes as needed for plug welds along the flange if resistance welds were used at the factory.

Imitate the proper folds of the outer skin with a piece of paper at the door's corners. Folding the edges of a piece of paper over the door will give you a perfect template for trimming the metal skin.

This is an example of a clean corner at the bottom of a repaired door skin.

After you've shaped your panel, lay it over the original door and trace the outer edge so you'll know where to bend the flanges. We ran this panel through the bead roller to establish bend lines for the flanges. Here we are checking the fit before we get too far along.

The Merc door is atypical in the way the front edge of the door has an extra step. Someone did quite a bit of welding on the back side of this flange at some point, so we thought better of attempting to remove it. Instead, we moved our weld seam to the nearest expanse of flat metal.

Don't forget that you can always shrink or stretch flanges as needed to fine tune the overall shape of a panel.

## LAY THE SKIN

Once you have all of your flanges bent at 90 degrees, lay the new skin in place on the door to check its fit once more. If you've taken your time, you will most likely have a really nice-looking door skin that will be tempting to weld. Do not be too hasty to weld the new skin. Instead, check to make sure your gap is good for the welding process you've chosen. If the gap is too narrow for MIG welding, for example, the edges of the panels will butt up against one another when you weld them and create a hump. If this happens, simply cut the weld across the hump, push the panels flat and reweld. You may have to do this more than once. Be patient and persevere. When TIG welding, gently hammer each small section of welds flat against a dolly to counteract shrinking. I have noticed with TIG welds that the metal gives perceptibly when you strike it on a dolly. There is a distinct just-right blow that feels good and the weld bead comes out perfectly flat. I realize that describing this sensation is a little like trying to describe the smell of a perfect chocolate chip cookie, but students have confirmed this sensation with me, so there must be some validity to it.

Regardless of the welding process you use, the first step is to establish a series of tack welds approximately ¾ inch apart across the long horizontal top seam of the door skin before welding the seam completely. After you've secured the skin with tack welds, fold the edges about 90 percent of the way over and test fit the door on the car. This trial fit is your last chance to make sure the door will be correct before you weld. If you are satisfied with the fit, remove the door and finish the long weld seam. Finish folding over the edges and plug weld them or weld them in the manner they were secured from the factory. Your new door skin and the original part of the door skin must stay in the same plane at all times during welding. The slightest discrepancy in alignment between the old and new metal will be impossible to correct by grinding without overthinning the metal. Sometimes clamping the seam with Vise-Grips is necessary or at least helpful to keep the new skin in perfect alignment with the original metal. If clamps are used, however, do not allow their weight to distort a panel that is soft from heat. At the outer edges of the door skin where it passes over the support structure beneath the skin, I recommend sticking a thin piece of copper between the outer skin and the inner structure. If you fail to do this, you will most likely weld the skin to the support beneath. A low spot will result that will be impossible to remove. By backing the weld with a piece of copper, you will allow the new skin to float free of the door support structure, as did the original. The skin will behave as one piece with the original part of the skin, it will be easier to bend the flanges along the edges, and the weld seam will be much easier to finish satisfactorily.

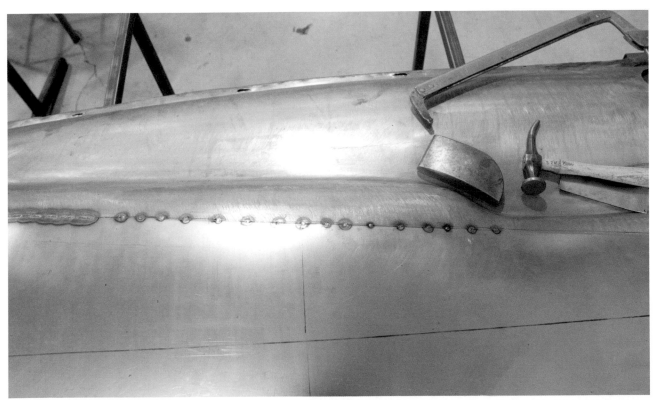

When the fit of your new skin is perfect, make a few tack welds in the center and work your way out to the sides, checking to be sure that everything is in perfect alignment the entire time. Hammer all tacks and eventually all welds against a dolly as you go to counteract shrinking.

Hammer all the door skin edges flat against a table or dolly, but do not get carried away; excessive bashing will introduce more curvature to the edges of the door. Crimp the edges only as much as the factory door edges. Do not hammer them into a knife-edge or they will look terrible when the door is open. Hit the flange high rather than close to the bend, and it will bend smoothly. Keeping the radius of the bend soft will allow you to tweak where the final edge ends up if you aren't sure, but you must support the exterior side of skin firmly with a dolly if you dare to hammer directly on the edge. If you fail to do this, you will make the exterior side bulge. Weld the skin at the lower corners. Grind your welds smooth, but do not overgrind. Proceed with caution so that you can even out highs and lows with the hammer and dolly. If you've done a good job of keeping the panels lined up, mysterious bumps that do not respond to gentle hammering will be from the panels butting into one another. Cut across the bumps with a cut-off wheel as instructed above and reweld, or heat-shrink the bumps if there isn't too much weld bead there. If the door will sit for any period of time, it is a good idea to spray it with self-etching primer to seal the metal.

## TROUBLESHOOTING DOOR GAPS

Another door issue worth addressing is the problem of large or uneven door gaps. We consumers have become accustomed to superb quality control and excellent panel fitment on today's cars. Consequently, the factory fit and finish of the cars of yesteryear at times leaves something to be desired to our persnickety eye. Sometimes the fitment of an old car's doors is unsatisfactory because of poor factory quality control, the installation of nonoriginal parts, prior damage, or structural deterioration. If there are frame issues, worn-out hinges, or damaged latch components, those are separate problems that must be addressed, but if everything is sound and a door gap is uneven or too wide, I have an easy solution. Weld a length of ⅛-inch-diameter welding rod to the door edge with the offensive gap. Using a MIG welder, alternate welds from one side of the wire to other to prevent the wire from pulling itself out of position. As you tack weld the wire in place, it is easily manipulated along the door's edge. Grind the face and back sides of the welds first, and then file the door edge with the largest, sharpest file you can find. Wear an insulated glove on your filing hand, and pull the file toward you along the door edge. By this method, you can obtain laser-straight gaps that rival those on any modern car.

Support the exterior of the door with a dolly or press it against the table as you hammer over the flange. An authentic Martin door skinning hammer works great for this, but an imported copy like the one shown will work. A plastic mallet is good for the initial hammering.

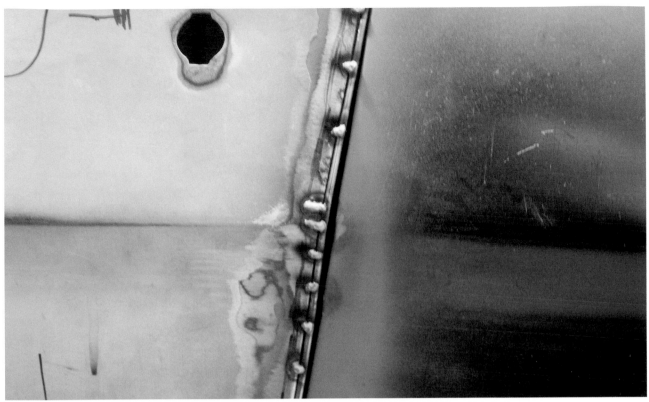

When faced with an unacceptably wide door gap, despite your best adjustment efforts, consider adding metal to the edge of the door. This looks crude, but it works well and is far easier than re-skinning an entire door. MIG weld a length of ⅛-inch welding rod along the door edge in the offending spot.

After careful grinding and filing, no one will be the wiser.

# Chapter 13
# Repairing Fenders, Hoods, and Trunk Lids

In this chapter I will examine some common maladies that afflict fenders, hoods, and trunk lids. Each car is different, of course, but patterns of corrosion and collision damage are surprisingly almost universal. Start with something small to build your skills and your confidence. Before long, very few projects will intimidate you.

## FENDER FUN

Fenders form part of the tender epidermis of our beloved automobiles. As such, they tend to receive a lot of damage, both from collisions and from the weather. I have presented here some examples of widespread fender problems with the hope that readers will be able to apply the techniques represented to repair their cars, whatever they may be.

Rust holes at the bottom of front fenders on the slab-sided cars of the 1950s and newer are extremely common

and easy to fix. Because patch panels vary in quality, it is difficult to know whether an aftermarket panel is a good option for your repair. I have illustrated a typical Ford Mustang patch panel to show how replacement panels compare to the original. In this case, the pressed body line in the reproduction piece resembled the original body line, and the fender lip along the wheel opening was somewhat close to the original, but it wasn't close enough simply to weld the new panel in place of the original. For $25, the patch panel was not a bad deal—and it would suffice for someone who did not feel confident to try making his or her own patch—but the finished result would not have had the quality we wanted. Furthermore, we had a rust hole about the size of a golf ball, so there was no reason to jeopardize the appearance of the body line and the fender lip.

No one would ever guess that the lower third of this Austin-Healey boot lid was a rusty, perforated disaster a few months prior. With the help of a hand-shrinking machine, fabricating a repair patch was not difficult.

The reproduction Mustang fender patch looks somewhat like the original, but we knew the different body line would be an eyesore that would annoy us if we used the panel.

Many front fenders develop rust where debris traps moisture behind the sheet metal.

Before you cut out the old damaged area of your original fender, determine if other repairs might be needed to the understructure and if there is a place for the weld seam with maximum accessibility from the back. Ask yourself if it makes more sense to fix the outer skin or the understructure first. I advise against trying to cut out both together because then you have fewer points of reference to help get everything back where it goes. Instead, repair either the support or the skin, and use the other piece to help you place your patch correctly. Fortunately, many front fenders needing this type of repair bend in only one direction in this area. In other words, this is not a crowned panel. Therefore, to recreate the curve in your patch panel you can bend the metal over your thigh, the arm of your sofa, or a nearby gas cylinder. You can also bend the metal using the English wheel by pulling down on the panel with light pressure between the wheels. When the curve is right,

the panel will sit contentedly on top of the original fender without rocking.

In the first Mustang example illustrated, we made the patch by bending, and then we traced around it and cut out the damage. We bead-blasted the support structure, taped off the edges of the panel where welding would take place, and coated the support structure with weld-thru primer. With the patch situated and clamped in the perfect position for welding, we traced the rear edge of the fender support structure onto the patch. We rolled the panel through the bead roller to create a bend line for the flange that forms the rear edge of the repair. We bent the flange to 90 degrees using a hammer and dolly, hand shrinking as needed to account for the reduction in length of the metal along the flange. When we were absolutely certain that the repair fit perfectly, we tack welded it in place and checked the fit on the car. Errors along panel edges are particularly noticeable.

No sophisticated tools are needed for this panel. We pressed a steel blank against a gas cylinder to get the correct contour.

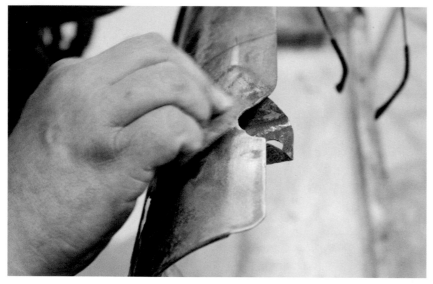

Hand-sanding spot welds is the best way to make them easy to see.

After the damaged portion of our panel was removed, we bead-blasted the area behind and sprayed it with weld-thru primer.

We used the bead roller to create the rear flange on this patch panel. It is ready to be welded.

Because we were satisfied with the fit, we welded the panel in but did not fold over the rear edge to facilitate the finishing of the weld bead. Without the panel crimped along its rear edge, we could still slide a thin spoon between the fender support and the outer skin to push on a low area next to the weld. With the weld bead finished to perfection, we hammered over the flange at the rear.

In the second illustrated version of the Mustang repair, I have marked areas with small pinholes that need to be welded. In addition, note that the length of the patch protrudes about a half an inch past the bottom of the original fender. Whenever you join two panels at an edge that needs to be trimmed to a specific dimension, leave one panel long until after welding. You will be able to weld right up to the edge without blowing a hole and you'll be able to trim the edge perfectly straight. Another phenomenon

worth mentioning is the tendency of panels to shrink along the weld seam. I have mentioned this already in the context of a straight seam, but the problem is exacerbated on a panel that is welded on two sides, as here, or even worse, if the panel is welded on all sides. The solution to shrinking, of course, is stretching, so hammer the weld seam against a dolly to stretch it back out. In addition, don't be surprised if you experience a lump in the panel where two seams come together at a right angle as here. A lump would also form if you hand shrunk a panel at the intersection of two edges that meet at 90 degrees. The metal bunches up like the waistband of your sweat pants when you pull the drawstring around your waist. Once forced up by shrinking nearby, however, the bulged metal will not return to flatness when you stretch the bead back out without heat-shrinking once directly on the bulge.

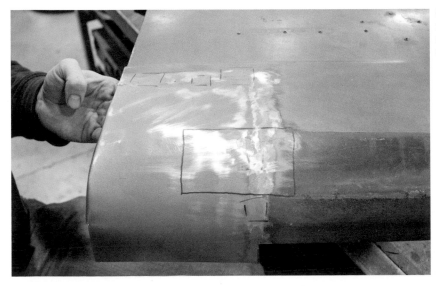

Scrutinize your welds for pinholes or discontinuities. I have marked some areas on this panel that need to be welded and/or bumped level. Notice that the patch is left long to facilitate perfect trimming once the welds are ground.

The blue line indicates the border of this freshly completed repair. If you look carefully, you can see the two faint blue spots on either side of the 90 degree angle in the patch where heat-shrinking was needed.

When we were satisfied with the repair to the Mustang fender outer skin, we cut off the corroded support behind and welded in an aftermarket replacement. The shape of the replacement was quite different from the original, but we did not have to be too concerned because we knew that the outer skin was correct. We just lined up the new piece so that it harmoniously fit our newly repaired skin and welded it in place.

Lance Butler's 1950 Ford had the same rust problem at the bottom of the front fenders as our Mustangs had, but the damage was worse. I could not ask for a better case to illustrate the fabrication of one of the bizarrely shaped fender support structures that often need to be replaced along with the outer skin. When faced with the prospect of trying to fabricate one of these confounding multisurfaced panels, break it down into manageable pieces that you can understand. If there are right angles involved, you can use the shrinker/stretcher machine to create curves. An edge with a soft radius can usually be replicated by hammering your metal over a piece of pipe of the appropriate diameter. A TIG weld will behave just like the parent metal if you need to weld two pieces of metal right on their edges to duplicate a particularly troublesome shape. In the case of Lance's fenders, the support structures were a grisly mess, so he broke the shapes down into a series of bent and bead-rolled pieces that could be welded together to form the correct shapes.

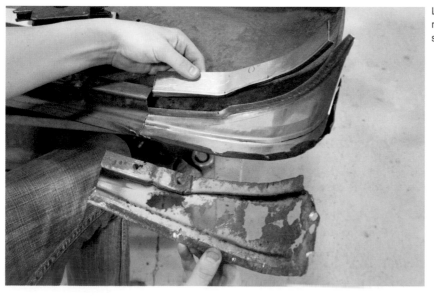

Lance Butler did a great job of combining bending, bead rolling, and shrinking to recreate the complex inner support for this front fender.

Notice that in the case of the Mustang fenders, repairing the outer skin first made sense. There was enough support structure behind to help us place the new pieces accurately. With Lance's Ford, however, the disastrous condition of the inner supports meant that they should be repaired first. Knowing that he could rely on the corroded but accurate outer skin to check the shape of his new panels, Lance repaired the inner supports without disturbing the outer skins.

Lance replicated the curve in his fender's outer skin by pulling down evenly on his panel as he wheeled it in the English wheel. Then he traced his panel onto the fender and cut out the damaged area. As with door skins, determining the precise edge of the new panel can be done by tracing the edge of the original panel on the replacement and bending the flange just inside that line or by tracing the inner support structure if it is really accurate, but you won't know the latter's accuracy as long as the outer skin is in place. As we did with the Mustang fenders, Lance bent the flange along the rear edge, sprayed the surfaces that would be inaccessible with weld-thru primer, tacked the patch in place, and checked the fit of the fender on the car. When he was happy with the fit, he welded the panel in and finished out the weld bead.

As you prepare to weld in a repair panel, there should be no question in your mind that its alignment is correct. Fit seldom improves during welding.

With all of the welding and metal finishing complete, Lance has a repair he can be proud of.

## '57 CHEVY FENDER REPAIR

Brandon Wight's 1957 Chevy took our fender problem to the next level of difficulty. First he had to repair the lower front fender and support structure for the right side. Then Brandon had to recreate the same missing elements for the left front fender using the right fender for reference. This is a classic old car problem—one side is bad, the other side is missing.

Brandon's repair of the first fender somewhat duplicates material we've already covered, so I will concentrate on the second fender. There might be a decent fender patch available for this popular car, but Brandon did not want to miss out on the terrific learning opportunity his car presented, so he was determined to make everything himself. The contour of the fender from top to bottom was a simple curve, fortunately, rather than a compound curve, so Brandon was able to make a vertical template by tracing along the rear edge of his good fender. Brandon's fender was made more challenging than the other fenders we've looked at by a complicated bead that runs along the edge of the wheel opening. To record this shape, Brandon made a series of templates running lengthwise down the good fender, as well as a template of the good fender's outer contour. He measured down from a string of factory trim holes drilled in both front fenders so that his templates would match from one fender to the other.

Brandon's cardboard template made from the opposite fender shows the desperate state of his right fender.

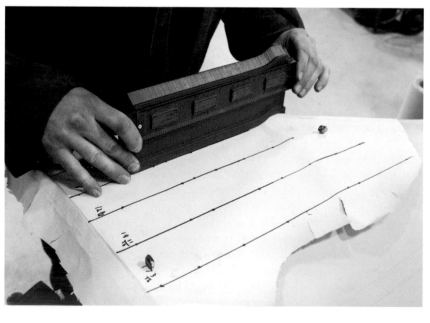

Unperturbed, Brandon uses a profile gauge to plot out a series of templates on his good fender that he can use on his bad fender.

All of Brandon's wheeling on this patch will be in a vertical motion because this patch is not a compound curve except on its edge.

To bring about the outward sweep of the fender near the wheel opening, Brandon hammers the patch against a T-stake with a plastic hammer.

Using his good fender for reference, Brandon made a cardboard template for his sheet metal blank and wheeled it into shape. He decided to split up the bead along the wheel opening because it would be almost impossible to make otherwise. To start the transition from the fender to the bead, Brandon flipped the panel over and hammered it against a stake along the edge that needed to curl up. Because this curved lip opposed the curve of the fender, a little stretching was needed or the lip would have pulled the shape out of the curve. On the other hand, too much stretching would have exaggerated the main curve of the fender. As he bent the transition up, Brandon checked the panel's shape against the vertical template.

Once the transition area along the wheel opening was bent, Brandon creased the outer edge of the panel in the English wheel using the flattest anvil wheel. The crease established a border for a narrow flat zone adjacent to the curved transition area. Brandon determined that the flat zone was the perfect place for a weld seam between the curved transition and the beaded outer edge of the wheel arch. With the panel in position on the bad fender, Brandon traced the rear edge and extrapolated the missing information for part of the contour from his templates.

With the fender lip started, Brandon tacked his new panel in place and checked the fit on the car. Satisfied with the fit, he welded the semicomplete panel in place and finished out the weld seam. The outer bead that runs along the edge of the fender lip was made out of a long strip of metal that began as an L shape and then became an L with a crease down the center of its widest leg. Brandon added another 90 degree bend opposite the first and stretched the edges until the piece followed the curve of the wheel opening. Brandon welded the last piece in place and finished the weld bead.

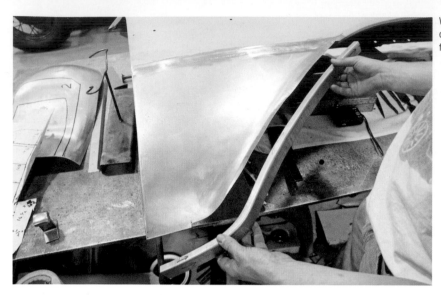

With the main patch welded in, Brandon checks the contour of the panel he has made to recreate the fender lip.

This close-up view of the fender lip shows how the fender lip is made and where it is welded. This scheme was a very successful approach to this problem.

Whether viewed up close or from a distance, the repaired area is visually indistinguishable from the original panel.

## MODEL A FENDER REPAIR

If you anticipate working on cars of the 1920s or 1930s, or even some British cars of the 1950s, I guarantee that you will have the opportunity to repair a fender that has been damaged across the beaded wire edge that runs along the outside of these fenders. I sometimes wonder whether fenders on early cars were used like primitive curb feelers—they were used to locate lamp posts, telephone poles, fences, trash cans, concrete buggy steps, nearby cars, and sundry other stationary and semimovable objects. By itself, repairing a dented fender edge does not sound overly daunting. Most of the difficulty in restoring a fender of this type emanates from the misguided repair attempts of previous owners and friends of previous owners who fancied themselves amateur body men. Sometimes, if the exterior paint still exists on one of these panels, you might not even realize the fender has been repaired because some craftsman of yesteryear has concealed the damage under a thick, artfully smoothed scab of plastic filler. If you look along the inside of one of these fenders, however, you are sure to see what looks like a painted mud dauber's nest—this is an old repair. As your mind fumbles for an explanation as to what the mysterious lump could possibly be made of—hardened cottage cheese, papier-mâché, popcorn ceiling texture—you realize that whatever it is, it's permanent. More often than not, it is a glob of brazing; sometimes it is a glob of brazing augmented with a hunk of steel rebar for additional support. You have two choices: 1) look behind you, make sure no one saw you discover this catastrophe, and get on with your life as if nothing happened; or 2) repair this mess. If you decide to attempt the repair, I'll describe the best way I have found to fix a damaged fender successfully.

First, resist the temptation to cut out the damage. Although its existence may annoy you like a wart on the end of your nose, do your best to bump the outer contour of your fender back into shape. The closer the contour of your panel is to its original shape, the easier it will be to fabricate a patch. When making any metal repair you should try to fabricate as much of any new panels as possible before cutting out the old metal. Use the original metal like a guide in shaping your new panel. If you have any doubt about how to shape your patch, lay a piece of paper over the anticipated repair and see what it tells you. Wrinkles mean the metal must be shrunk. If you must cut or tear your paper to fit, the metal must be stretched. If your panel has any inner support structure that helps hold it to the car, install the support if possible to help stiffen the piece. You do not want anything moving out of place once you cut out the old damaged metal. In addition, the larger the patch, the less freedom you have to roll the bead without distorting the patch. If your patch climbs up several inches on your fender, you might have to consider making the bead as a separate patch, extending only about an inch or two past the upper edge of the bead.

A suitable patch for a highly crowned fender of the type illustrated here may easily be made by stretching a crown in a panel with an English wheel or with a hammer on a shot bag. An alternative method is to create a crown by shrinking around the edge of a panel with a hand shrinker. Keep shaping until your patch will sit on top of the original panel without rocking. A simple rule of thumb is to stretch your patch wherever it touches the original panel until it touches everywhere without tension. For example, if your patch rocks on its center, stretch the center to raise the center, thereby creating a deep pocket, like a baseball catcher's mitt, that gently envelopes the portion of the original fender on which it sits. Persevere and do not get discouraged. Often times, learning to fabricate things from sheet metal can be a little like a bar room brawl—noise and ugliness precede a resolution, but the tenacious prevail.

The example repair I present here works well for 1928 through 1932 Ford fenders. The 1928–1929 fenders are narrower than the later fenders, but the repair procedure is exactly the same. To create projects for students, I plasma cut original fenders across the fender bead and cut up several inches into the panel to simulate a long crack. This repair is most easily accomplished using two patches—one up high over most of the fender's crown and a second patch closer to the bead. For an alternative way to repair this type of fender, consult John Glover's *The Practical Sheet Metal Worker*, in which Glover flattens the bead, patches the flat metal, and finishes by rolling the bead.

If your repair will consist of two patches—one on the fender proper and one on the fender bead—leave the bead undisturbed until your fender patch is welded in on three sides. The bead will help stabilize the fender and will facilitate the welding of the first patch. The most important thing to keep in mind when you weld in your patch is that it absolutely must stay in the same plane as the surrounding metal. If at any point the patch climbs up or down, you should stop and correct the problem. Cut out any tack welds as needed to realign the metal. With TIG welds and torch-welds, promptly hammer each weld against a dolly to counteract shrinking along the weld seam as you progress. At every point throughout the repair you should be able to run your hand over the welds and confirm that all of your metal is in the same plane.

For the typical old Ford fender, start by repairing the highest part of the fender, rather than the bead. Cut out a blank that is larger than you need because it will be easier to hold onto while wheeling. Select an anvil wheel with a slight crown; it will probably be the second to the flattest wheel you own. Run the panel lengthwise through the English wheel in even strokes close together. Remember that the flatter anvil wheels typically create more shape in the direction that the panel is passed back and forth through the wheel, rather than to the sides. Wheel the panel and check its shape against the original fender until the patch will sit on the fender without rocking front to back. The sides, or at least the side nearest

We start the large upper Model A patch by wheeling lengthwise with a semiflat anvil wheel. The panel will curve predominantly front to back and we'll generate a little shape across the piece.

We push the panel all the way into the hand-shrinker's jaws and shrink along the edge adjacent to the bead. The more money you have spent on your shrinker at this point, the better your results will be. Our results are satisfactory.

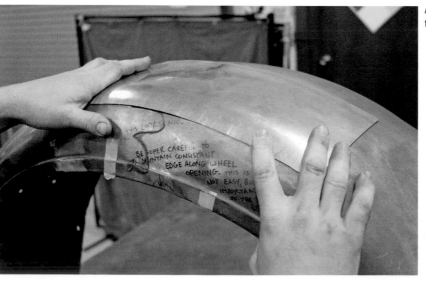

After wheeling lengthwise and shrinking the outside edge, the first patch fits well.

the bead, will still be high at this point. To encourage the side of the patch nearest the bead to curl over tightly against the fender, shrink along its edge with the hand shrinker. If the transition between the shrunk outer edge and rest of the panel is too abrupt, wheel across it.

When you are satisfied with the first patch, go ahead and trace around it, cut out the original damaged portion of the fender you seek to replace with this patch. Weld in the new panel on the three sides excluding the side nearest the fender bead. Remember, welding one patch in at a time keeps the fender stable so that things don't spring out of place. Decide what type of welding process you will use to attach your patch and cut out the old metal leaving enough metal to establish the proper gap for welding. For TIG welding, the patch can butt against the adjoining metal around the patch's perimeter. For MIG welding, leave a 0.040-inch gap, which is just a hair more than the thickness of the metal in most cases.

To create the patch along the fender bead, cut a piece of metal a little larger than you need and shrink the top edge to encourage it to curve over. Too much shrinking will cause the metal to bunch up, however, and it will sit high, so balance your shrinking with some bending against a stake with a rawhide mallet. The reason you cannot just bend the top of the patch over is that the distance occupied by the curve is shorter than the length of the flat panel. To get the metal to occupy the shorter distance, therefore, you must do some shrinking. When the patch will sit happily against the fender but for the interference of the bead, you are ready to mark and roll a bead along the bottom edge of the patch. Run a few test pieces of steel through your bead roller to find the optimum pressure and distance to replicate your original bead with your dies. If a bead roller is not available, it is possible to create a homemade die that works satisfactorily. Cut a piece of flat steel of the appropriate thickness to create the offset in the fender, drill holes through the steel, and weld it to another flat piece of steel. Grind all the welds flat. This process creates a die over which you can clamp a flat piece of sheet metal and chase out the correct bead with a set of dull chisels.

We begin the second patch, the one that will have the bead, by shrinking along its top edge where it will join the first patch.

Having shrunk the top edge of the patch slightly, we hammer it over a stake with a rawhide mallet to induce a soft curve that will transition smoothly into the first patch.

If you lack a bead roller, clamp your blank between two pieces of metal and chase the bead shape out with a punch. For curves, cut the curved portion of your die out and plug weld it to another flat sheet of steel. This will make clamping it down easier.

The easiest way to copy the arc of the original bead is to mark the edges of the patch where it overlaps the original bead. Lay the patch behind the fender and trace the curve of the fender edge between your two reference points. This shouldn't really work because the curve of the top of the bead must be a larger radius than the curve of the bottom of the bead, but I've done this many times and it always seems to look believable.

Exercise great care when cutting out the old metal along the fender bead so that you do as little damage as possible to the wire inside the wire edge. If the wire needs to be cut out and repaired, use 1/8-inch-diameter welding rod, or mild steel rod of the same diameter as the original wire. One tip I have discovered that makes this repair much easier is to avoid overlapping the sheet metal repair with the wire edge repair. If you attempt to weld the wire and the outer sheet metal in one fell swoop you will not be able to get a smooth, imperceptible weld seam. Instead, repair the wire, grind the welds smooth, and repair the sheet metal at least 1/2-inch or an inch beyond the repair to the wire. Make your patch, create the bead by whatever means necessary, and prepare the edge of the patch to wrap around the wire as demonstrated in the chapter on hand forming. Coat the metal that will enclose the wire with weld-thru primer to prevent corrosion. Weld the patch in almost completely before wrapping the outer edge around the wire. Wrap and crimp the edge of the repair, make sure that the repair does not interrupt the flow of the fender when viewed from several feet away, and then finish welding in the

We superimpose the second patch over the fender and mark where the bead crosses the panel on each side.

We place the panel behind the fender and trace the fender's contour onto the panel to use as a guide when rolling the bead, even though common sense suggests that this shouldn't work.

After bead rolling, the repair looks convincing. The next steps are to create a bottom flange, weld in the patch, roll the wire edge, and grind all the welds smooth. See the beginning shaping chapter for a finished example of this repair.

patch where it wraps around the wire. When grinding your welds, remember that only the weld bead needs to be ground. Consider applying a strip of duct tape over the fender bead on either side of the weld to protect it from inadvertent grinder marks. Any nicks left in the fender bead will be noticeable, so grind out your welds with caution. If your repair is MIG welded, start your grinding with a weld grinding wheel in a pneumatic cut-off tool or use an electric 3-inch grinder. Grind the bead until it almost flush with the surrounding metal, then move to a 3- or 5-inch grinding disc with something coarse like 24 or 36 grit. Then switch to a smoother paper like 80, followed by 150. Cover the raw metal with some rattle can self-etching primer to prevent rusting until you have time to prime and paint the entire fender.

## HOODS AND TRUNKS

Headaches caused by hoods and trunk lids come in the form of collision damage, corrosion, and poor fit generally. Hoods seem less prone to rust than trunk lids, in my experience, but they certainly receive their fair share of collision damage. The best way to repair collision damage along the front edge of a typical hood on a 1950s or newer car is to drill out the spot welds that secure the outer skin to the inner support structure. Cut out as much of the inner structure as necessary to get to the damaged outer skin. If the outer skin is severely folded up, tension applied from a hand-held winch—a *come along*—is a great way to straighten out the sheet metal if a lot of force is needed. Attaching your puller to the sheet metal might take a little creativity because you need to spread the load out so as not to create more damage than you are removing, but the good news is that the outer skin moves a lot easier now that the inner structure is gone. Take a short strip of angle iron, clamp it several times along its length to the edge of the hood, and weld on a few large washers a few inches apart. Pull from the washers. A slide hammer with a pulling hook on the business end might generate enough force to move your metal if only the outer skin is concerned. Also, you need to anchor the hood so that it is immoveable; don't pull it on its hinges. We have a colossal cast-iron table for such purposes, but I would never suggest acquiring such a monstrosity unless you're looking for an excuse never to move from where you now live. An easier solution is to lay the hood on the garage floor or workbench, chain the hinge mounts to the wall, and pull from the opposite wall. For stubborn dents, apply a few judicious raps with a big rubber mallet on the recalcitrant creases while the hood is under tension. Once the largest dents are out, you can straighten the hood the rest of the way using the techniques outlined in the section of this book on dent repair.

Prewar hoods tend to be flatter than later hoods, so they require a little different strategy. We recently made a hood side for a 1928 Franklin, because there was no amount of straightening that could save the original, but at least we could measure the original to copy the dimensions. The hood side was mostly flat, but it had multiple bends along its bottom edge and a typical 1920s- to 1930s-style hood hinge along the top edge. With a Pullmax or other similar machine, I suspect one could fabricate some dies to recreate this hood hinge. It is very much like a wire-wrapped edge with intermittent rectangles nibbled out of it. We decided to separate the complex multifold lower portion of the panel from the rest of the hood side because we did not see how we would be able to do the entire piece with the equipment at our disposal. Furthermore, by adding the lower portion separately, we were able to fine-tune the final width of the hood with less chance of a mistake. We were doubtful that we would be able to make the many bends, roll the hinge, and still end up with the exact width of the original, which was narrower at one end than the other, by the way.

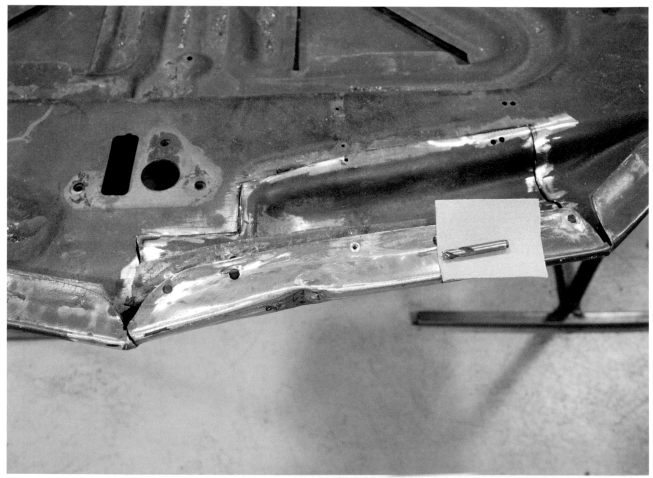

This Chevelle hood is about to have its inner support structure removed behind the damaged area. We sanded the spot welds by hand, drilled them out with the spot weld bit, and cut generously around the damage with a cut off wheel. This dent will be easy to pull without support behind it.

We are in the process of cutting reliefs for hood hinge clearance in our Franklin hood blank.

We spent considerable time testing different techniques and blank dimensions before we perfected our method for the fabricating the hinge. The first key to success was bending the small metal tabs in the hinge over a small homemade die that could be slid along the bead at the top edge of the hood side. The second important decision was Chris Paulsen's suggestion to wrap the tabs around a mild steel rod that was 0.007 inch larger than the rod that runs through the finished hinge. The larger rod, obtained from MSC Industrial Supply (www.mscdirect.com), worked beautifully for achieving a good fit around the correct rod without binding. After measuring repeatedly, we welded on the multifold panel along the bottom edge of the hood side. Some shrinkage along the weld concerned us because it caused the entire hood side to oil can. We ran the welded edge of the hood side through the English wheel with the flattest anvil wheel installed several times, flipping the hood over with each pass. By gradually stretching the metal along the weld, the tension was released and the oil can went away.

Mikhail Perez and Chad Ediger made this die to slide along the bead at the top of the hood side. The die conforms to the sheet metal in this area and allows us to bend each hinge loop perfectly without distorting the surrounding metal.

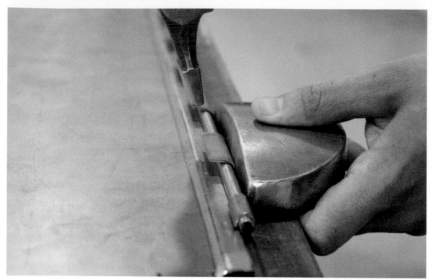

After the initial forming of each hinge loop, a hammer is brought into play to crimp the metal down supported by a dolly.

The final step with regard to the hinge is to crimp each loop around an oversized rod with the hand-wiring tool.

When we welded the multifold panel along the bottom of our hood side, shrinkage turned the panel into a giant oil can. Fortunately, we wheeled the panel along the weld seam several times, flipping the panel over each time, and the tension went away.

Trunk lids, called *boot lids* in the United Kingdom, seem to collect water and rust from the inside out. When faced with a rust-perforated trunk lid, evaluate the inner structure and form an action plan. Will you first address the outer skin or the inner structure? On Austin-Healey boot lids, for example, I typically drilled out the spot welds retaining the lid's inner support, removed and repaired the inner support, then screwed the support back in place before addressing the outer skin. Sometimes I did it the other way around. By tackling one piece at a time, I always had something to rely on for checking fitment. The panels themselves were classic shrinker/stretcher projects, but satisfactory results would have been much more elusive had I cut out both damaged areas and attempted to repair them simultaneously.

I repaired the outer skin of the Austin-Healey boot lid. Now I'm about to address the inner support, which has had its spot welds drilled out.

The outer skin of a trunk or boot lid often needs a repair like this along its rear edge. All you need is a shrinker/stretcher machine.

Some of the more aggravating issues hoods and truck lids present are mysterious fitment problems. When panels are not original to a particular car, you can end up with enormous gaps or panels that are too big for their intended space. The picture of the Triumph TR3 bonnet documents my attempts to find the best fitting of three bad bonnets. Because I couldn't simultaneously compare them with one another, I took a series of pictures from several vantage points and pondered which issues I thought I could resolve. Bonnet No. 1 was the one I chose for the car. To improve its fit, I cut partially through the bonnet's understructure in a few places, pried against the skin to change the bonnet's curvature where necessary, and crawled into the engine bay from underneath the car to tack weld the cuts made earlier at strategic points. After reenacting this limbo ritual several times, I obtained acceptable panel fitment and welded up the bonnet's understructure.

Another comical instance of poor hood fitment arose with our '50 Mercury, a coupe that someone converted into a convertible. Among this car's many peculiarities, the cowl at the base of the left A-pillar had been filed through in a desperate attempt to make room for a hood that was too small. The gap all along the rear edge of the hood looked great except for one area on the left side. Unable to find another 1/16 inch of wiggle room in the hood, and unwilling to grind the edge of the hood for fear that it would be noticeable, we cut the edge of the cowl, pushed it back about 1/16 inch and tack welded the seam. Moving this edge in this manner would not have been possible if the metal had been overlapped in any way along the bend, but it was not. I don't recommend this kind of quasi-junkyard trickery in every situation, of course, but it is an extreme case that is worth keeping in the back of your mind in the event you find yourself in a similar situation and you are desperate for a solution.

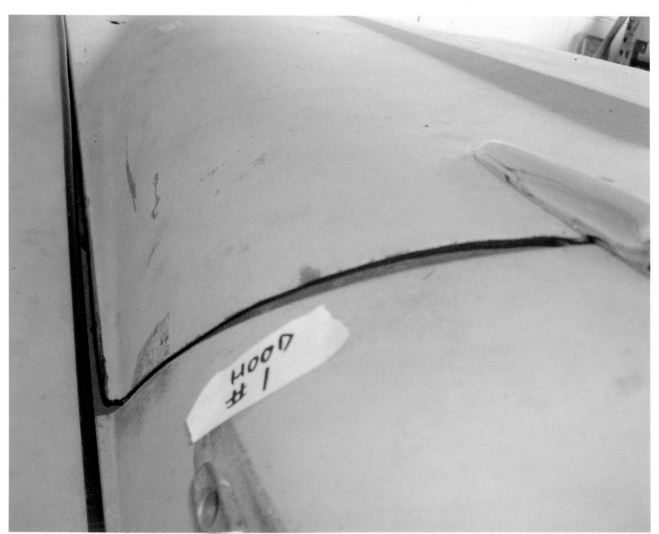

This bonnet was the best of three that I had for this car, and the fit was wretched. I cut the support structure underneath so that I could relax the curvature of this panel near the cowl, also known as the scuttle. When I got the bonnet fitting the best that I could, I welded up the various cuts I had made underneath.

This Mercury cowl has been filed through by some desperate soul in an attempt to create clearance for the hood that could not be made to fit. After exhausting all methods of stretching the hood, we made a couple of cuts in the cowl, pushed the edge back 1/16 inch, and achieved a clean gap against the hood.

Most of the time, poor panel fit can be rectified through careful adjustment, but I know I have encountered panels that were overtrimmed from the factory. I remember MG T-series cars with hand-trimmed bonnets that were slightly jagged along their rear edge. There are a few ways to approach this situation. You can leave the car alone because that is how it left the factory. Small idiosyncracies such as I have just described are part of the charm of old, largely handmade cars. I think of these details like the chalk signatures of women who signed their names on the backs of door panels and seat covers of these same cars—they are part of the car's history. If you are working for a customer, however, you must consider what his or her expectations are. The customer may appreciate your connoisseurship but he or she may rather have a straight hood. Some options would include gingerly filing the rough edge to even it out somewhat. You could remake the hood side, or you could weld on a strip that is wider than needed, grind the welds down, and then trim the edge to suit the owner. The Jaguar XK 120 illustrated had an unevenly trimmed bonnet and boot lid, and the decision was made to fix them. The only solution for undersized panels like these is to add metal in the form of a patch or multiple weld beads and regrind to the correct shape. The E-type boot lid, on the other hand, was the victim of substandard bodywork on the sheet metal surrounding the boot lid opening. The rear panels, which had been replaced after a collision, just did not line up with the boot lid. The solution was to remake the rear edge of the boot opening.

This Jaguar XK120 boot lid was overtrimmed from the factory. As the rest of the car started to look great, the deficient areas really stood out. I welded a few beads along the edge of the panel and filed it down so that the gap looks even.

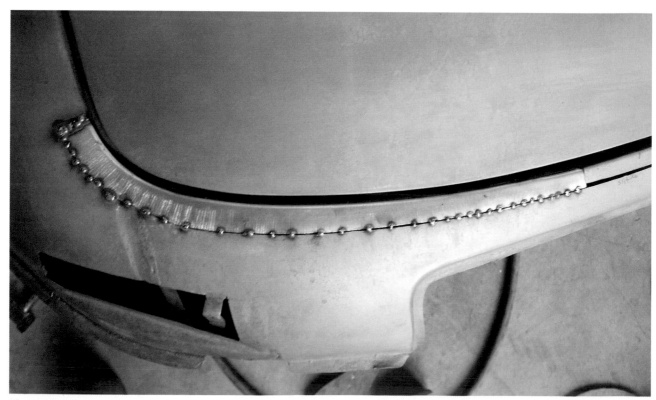

After several rear body panels were poorly installed following a collision, the fit of the boot lid was exceedingly poor on this Jaguar E-type. The expedient solution was to remake the edge of the boot opening in the questionable areas.

# Resources & Bibliography

## BIBLIOGRAPHY

Barton, Timothy Paul. *Metalshaping: The Lost Sheet Metal Machines*, vol. 1–4, Autofuturist.org, 2010.

Bonacci, Nick. *Aircraft Sheet Metal*, Jeppesen Sanderson, 1987.

Bridigum, Todd. *How to Weld*, Motorbooks, 2008.

Cowan, T. N. *Automotive Body Solder*, Motorbooks International, 1980.

Glover, John F. *The Practical Sheet Metal Worker*, John F. Glover, 2006.

Lilly, Jeff. *How to Restore Metal Auto Body Trim*, Motorbooks, 1997.

Sargent, Frank T. *The Key to Metal Bumping*, Fourth Edition, Martin Tool & Forge, 2006.

Sargent, Robert L. *Chilton's Mechanics' Handbook, Auto Body Sheet Metal Repair*, vol. 3, Chilton Book Company, 1981.

## RESOURCES

Broadway Blasting Services, 410 Broadway, Hanover, PA 17331 (717) 630-2260

http://frwebgate.access.gpo.gov/cgi-bin/getdoc.cgi?dbname=2009_register&docid=fr11au09-20.pdf Conservative acetylene flow recommendation.

http://stores.ebay.com/pro-shaper-tools-and-videos Wray Schelin's shrinking disc.

www.airgas.com/content/details.aspx?id=7000000000128 Comparison of gas consumption of oxy-fuel cutting tips.

www.alcotec.com All about strengths and weakness of aluminum alloys.

www.aluminum.org Aluminum filler alloy chart.

www.eastwood.com Automotive restoration tools.

www.faybutler.com Author and vendor of a book about die-making for Pullmax-style machines, plus many other resources.

www.fournierenterprises.com Metal-shaping tools, aluminum-welding goggles.

www.ppg.com Automotive paint supplies and metal preparation solutions.

www.solarflux.com Flux for welding stainless steel.

www.tinmantech.com Metal-shaping tools, aluminum-welding goggles.

www.vansantent.com Metal-shaping tools.

www.vwmc.com Victory White Metals Company, vendor of automotive body solder.

# Index

abrasive blasting, removing paint by, 213–214
acetylene, rate of flow, 80
acetylene cylinders
    about, 24–25
    capacity, 80
    fittings for regulator and hose, 28–30
    moving and storing, 25–26, 28
    obtaining/refilling, 25, 27–28
    opening, 26–27
    safety precautions, 25–27
    shutting down, 34
acid core solder, 65–66
*Aircraft Sheet Metal* (Bonacci), 78
alignment, maintaining, 41–42
aluminum
    annealing, 98
    brazing compared to torch welding, 62
    buffing, 239
    removing paint from, 213
    shaping using stumps, 115, 116–120
    thumbnail shrinking, 189–190
    TIG welding, 48, 56, 58–60, 253
    torch welding, 43–46
    unreceptive to welding, 60
    *See also* fender construction project
Aluminum Association website, 43
aluminum oxide, for blasting corroded steel, 214
Amco 22 aluminum welding flux, 43
annealing, 10, 98
antique fender modification project, 219–223
argon
    for MIG welding, 48
    for TIG welding, 55
Austin-Healey boot lid repair, 297
auto-darkening welding helmets, 47
*Automotive Body Solder* (Cowan), 72, 73

back purging, 60
backfires, 35
backhand welding, 37
Barton, Timothy Paul, 161
batteries, 5
bead rollers, 18
beading
    preparation of surface, 139
    process, 140–145
    with reciprocating machines, 166–169
bending, 95, 115, 135–136
Beverly shears, 15–16
blasting, removing paint by, 213–214
body hammers
    defined, 10
    types, 11
body solder
    adhesion of lead to sheet metal, 67
    advantages and disadvantages of, 67
    defined, 10
    lead sleds, 72–73
    metal percentages in, 66–67
    process, 67–71
    safety, 67, 68, 72
Bonacci, Nick, 78
bonnet repairs, 298–300
boot lid repair, 297
boots, 5
brass, 65, 239

brazing
    advantages and disadvantages of, 62
    process, 63–65
    rods, 62–63
Bridigum, Todd, 24
bucks
    defined, 10
    making wood, using Styrofoam models,
        183–186
buffing and polishing, 238–241, 242
bull's-eye picks, 10, 13–15
bumping hammers, 10, 11
Butler, Fay, 94–95, 161, 166
butt joints
    making on aluminum by TIG welding, 59
    making on steel by TIG welding, 58
    making with MIG welders, 52
    making with oxyacetylene torches, 40–42
    small floor repairs, 255–256

carbon dioxide, for MIG welding, 48
carbon steel, torch cutting, 80–81
carburizing flames, 31, 33
cast-iron and torch cutting an, 81
Caswell Plating, 239, 241
chemical dip stripping, 212
chemical paint strippers, 213
Chevy fender repair project, 286–288
chrome plating, 242, 243
clecos, defined, 10
clothing, safety. *See* protective clothing
cold-shrink, defined, 10
cold-working, defined, 10
combination hammers, 11
compasses, using, 147–148
compound curves, making using English wheels,
    128–131
conditioning treatments, 214, 215
copper
    buffing, 239
    safety precautions, 73
    silver solder and, 65
copying shapes, 146–149
corking tools, 10, 13
corrosion
    dissimilar metals and, 76
    priming to avoid, 212, 213, 215, 292
Cowan, T. N., 72, 73
crimping. *See* tucking
crisp lines, making using English wheels, 136–139
crowns, making using English wheels, 128–131
curves, making reverse, 169–173
custom pedal car project
    background of, 244–245
    exhaust system fabrication, 248–252
    grille fabrication, 252–253
    steering wheel fabrication, 253–254
    track nose fabrication, 245–248
    wheel arch fabrication, 252
cut lines, 274

definitions of terms, 10
dents, tools for small, 15
designs, cutting tips, 83
detail work, making crisp lines using English
    wheels, 136–139

dies, choosing, 162
dinging hammers, 10, 11
documentation, importance of, 8–9
dolly blocks, 10, 11–12
doors
    gap repairs, 278–279
    skin replacement project, 270–278
    straightening project, 229–234
drag, described, 85
Drake power hammer, 161–162, 169
driving tools, 10, 13
Durbin, Bill, 214
dzus fasteners, 10, 78, 247

Eastwood Tinning Butter, 66, 67, 68–69
Eastwood tools, using, 16–18
elastic limit, described, 10, 161
electrical connections, solder for, 67
electroplating, 241–243
English wheels
    about, 10, 19
    making bends using, 135–136
    making compound curves using, 128–131
    making crisp lines using, 136–139
    tips for using, 132–134
envelope (of flame), 31
ER specification, explained, 48
eyewear, safety, 5, 34

face shields, 34
Fairmount tools, 10
feather (of flame), 31
fender construction project
    final adjustments, 203, 206–207
    first steps, 188–190
    options, 187–188
    planishing, 203, 208–210
    step-by-step first section construction,
        190–201
    welding, 201–206
fenders
    Chevy repair project, 286–288
    dent repair project, 219–223
    Ford repair project, 284–285
    making shapes like, 169–173
    Model A repair project, 289–293
    Mustang repair projects, 281–284, 285
    patch panels, 280
files, 16
filing, described, 220
filler rods
    for aluminum, 43–44, 46
    for stainless steel, 60
    for steel, 38–40
    TIG welding and, 56–57, 58
flame colors (oxyacetylene welding), meanings
    of, 31–32
flange welds, 38
flexible patterns, 148
floorpan repairs, 255–257
floorpans
    large repairs, 256–257
    small repairs, 255–256
flux
    for brazing, 63
    for lead solders, 65

for TIG welding of stainless steel, 60–61
for torch-welding of aluminum, 43, 45
flux-core wire arc welders, upgrading for MIG
  welding, 49
Ford fender repair project, 284–285
Ford grille shell project, 173–180
Ford Model A rear quarter panel repair project,
  263–266
form/forming, defined, 95
Fournier Enterprises, 44
Franklin hood (1928) project, 294–296

gas welding. *See* oxyacetylene welding
Glover, John, 19, 289
gloves, 5, 34
gouging
  described, 88
  nozzle shield usage, 89
  process, 91–92, 93
grains of metal, action of, 216, 218
grille shell project, 173–180
grinding welds, 237–238
GTAW (gas tungsten arc welding).
  *See* TIG welding

hammerforms
  described, 10, 111, 112
  fabricating wheel arches, 252
  using, 111–114
hammers, 10–12, 97, 99
  *See also* specific types of hammers
hand-operated shrinkers/stretchers (Eastwood
  tools), 16–18
heat-shrinking
  described, 10, 102
  example, 107–108
  steel, 223–227
helium for TIG welding, 55
holes, filling, 54
hood repairs, 293–296, 298–300
*How to Restore Metal Auto Body Trim* (Lilly), 239
*How to Weld* (Bridigum), 24
hydrogen for welding aluminum, 43
hydrostatic test dates, 25
Hypertherm 45 (for plasma cutting), 89–90

jack stands, 7

kerf, described, 81
*The Key to Metal Bumping* (Sargent), 216
Knight, Scott, 235

labeling, importance of, 8–9
lead body solder
  adhesion of lead to sheet metal, 67
  advantages and disadvantages of, 67
  lead sleds, 72–73
  metal percentages in, 66–67
  process, 67–71
  safety, 67, 68, 72
lead-free solder, 73–74
lead sleds, 72–73
leaf spring mounts, 7
Lilly, Jeff, 239
linear stretching dies/hammers, 10, 97

mallets, 15
Mar-Hyde Tal-Strip II, 213
Martin tools, 10
metal inert gas welding. *See* MIG welding
metal memory, 214, 216

*Metalshaping: The Lost Sheet Metal Machines*
  (Barton), 161
MIG welding
  advantages and disadvantages of, 47
  butt joints, 52
  cut line, 274
  equipment, 47, 48–50, 54
  gases for, 48
  grinding beads, 238
  machine set-up, 50, 51
  patching tips, 291, 293
  process, 47–48
  running beads, 50, 52–54
  safety, 47
  wire selection, 48
mini-power hammers, 161–162
Model A fender repair project, 289–293
MSC Industrial Supply, 295
Mustang fender repair projects, 281–284, 285
Mustang wheel arch repair project, 267–269

neutral flames, 31, 33
nickel plating, 242, 243
non-ferrous metals, torch cutting, 81
nozzle shields, 89

on and off dolly, defined, 10
oxidation, removing, 212–214, 215
oxidizing flames, 31, 33
oxyacetylene cutting torches. *See* torches
  (oxyacetylene cutting)
oxyacetylene welding
  aluminum, 43–46
  butt joints, 40–42
  clothing for, 34
  equipment set-up, 28–30
  with filler rods, 38–40
  hydrostatic testing of tanks, 25
  lighting and adjusting torch, 30–33
  manipulating torch, 35–37
  metal for practicing, 34
  running beads, 35, 37
  shutting down torch, 34
  supplies, 24–28
  torch safety, 24, 34–35
  work area, 34
  *See also* acetylene cylinders; oxygen cylinders
oxygen cylinders
  about, 24–25
  fittings for regulator and hose, 28–30
  moving and storing, 25–26
  obtaining/refilling, 25
  opening, 26
  safety precautions, 25–27
  shutting down, 34

paint, removing old, 212–214, 215
parts, keeping track of, 8–9
pasty ranges, described, 67
patches, making, 197–198
pedal car project. *See* custom pedal car project
picking, described, 220
piercing, 91, 92
planish/planishing hammers, 168
  described, 10
  making reverse curves with, 169–173
planning, 6–7
plasma cutters
  cutting ability of basic, 79
  parts of, 88–89
  piercing vs. cutting capacity, 91

plasma cutting
  advantages of, 87–88
  air pressure requirements, 89, 90
  process, 87, 90–91
  safety, 88
  using Hypertherm 45, 89–90
plastic blasting media, 214
Plumb tools, 10
pneumatic combination hole punch and
  flanging tools (Eastwood tools), 16–18
polishing and buffing, 238–241, 242
pop rivets, 75
power hammers, 161–162
*The Practical Sheet Metal Worker* (Glover), 289
priming, to avoid corrosion, 212, 213, 215, 292
profile gauges, using, 146–147
projects
  antique fender modification, 219–223
  Chevy fender, 286–288
  door straightening, 229–234
  fender dent, 219–223
  fender shapes, 169–173
  Ford fender, 284–285
  Ford grille shell, 173–180
  Ford Model A rear quarter panel, 263–266
  Franklin hood (1928), 294–296
  Model A fender, 289–293
  Mustang fender, 281–284, 285
  Mustang wheel arch, 267–269
  shape copying, 121–128, 146–149
  shot bags, 20–23
  skin replacement, 270–278
  small gas tank, 150–159
  *See* custom pedal car project; fender
    construction project
protective clothing, 5
  for lead soldering, 67, 68
  for MIG welding, 47
  for torch cutting, 84
  for torch welding, 34, 35
Proto tools, 10
prying tools, 13–15
puckers. *See* tucking
puddles, creating, 34–35, 36
  with filler rods, 40
pure movement, described, 161

quarter panel repairs
  access problems, 261–262
  Ford Model A rear, 263–266
  Mustang wheel arch, 267–269
  planning, 260–261, 262
  using ready-made panels, 259–260

reducing flames, 31, 33
resistance welders, 227–228
respirators, 67, 72
riveting
  making corrections, 78
  process, 76–78
  rivet selection, 75–76
rocker panel repairs, 257–259
rotisseries, 7

safety
  buffing, 240
  clothing, 5
  lead soldering, 67, 68, 72
  MIG welding, 47
  oxyacetylene torch cutting, 83, 84
  oxygen and acetylene cylinders, 25–27

plasma cutting, 88
sanding, 212
torch welding, 24
torch welding aluminum, 44
torch welding steel, 34–35
tungsten electrodes, 56
working with copper, 73
sand bags, 121
sanding, dangers of, 212
Sargent, Frank T., 216
Schelin, Wray, 235
shape/shaping
copying, 121–128, 146–149
defined, 94–95
thickness and surface area relationship, 94–97
shears, 15–16
sheet metal brakes, 15–16
shoes, 5
short-circuit transfer gas metal arc welding. *See*
MIG welding
shot bags
making, 20–23
using for shaping, 120–128
shrink, defined, 10
shrinking
by cold tucking, 100–107
counteracting, 289
by heat tucking, 102–104, 106, 107–108
straightening and, 217–219
with thumbnail dies, 162–165
torch-shrinking steel, 223–227
using planishing hammers, 169–173
using resistance welders, 227–228
using shrinking discs, 235–237
using simple mechanical shrinkers, 109–111
using stud guns, 227, 228–229
using stumps, 115–116
shrinking discs, 235–237
shrinking hammers, 11
silver soldering, 65–66
slappers (slapping spoons), 10, 13
slip rolls, 19–20
small gas tank project, 150–159
smoothing, 115
Solar flux, 60–61
soldering
lead-free, 73–74
silver, 65–66
soft, 65–66
*See also* lead body solder
solid rivets, 75
spoons, 10, 12–13
stainless steel
buffing, 239
TIG welding, 60–61
torch cutting and, 81
stakes, defined, 10
steel
aluminum oxide for blasting corroded, 214
shrinking, 103–104, 106, 107–108
TIG welding, 48
torch cutting, 80–81
torch shrinking, 223–227
torch-welding
butt joints, 40–42
with filler rods, 38–40
flange welds without filler rods, 38
manipulating torches, 35–37
practicing, 34
safety when, 34–35
straight lines, cutting tips, 83

straightening
metal memory and, 214, 216–217
shrinking and, 217–219
stretching and, 217–219
torch shrinking steel, 223–227
work-hardening and, 214, 216
stretching
defined, 10
straightening and, 217–219
using English wheels, 128–131
using mini-power hammers, 166
using planishing hammers, 169–173
using shot bags, 120–128
using shrinking hammers, 11
using simple mechanical stretchers, 109–111
using stumps, 115, 116–120
stretching metal, 97–99
stud guns, 227, 228–229
stumps
described, 114–115
shrinking using, 115–116
stretching using, 115, 116–120
Styrofaom models, using, 183–186
surface finishing
buffing and polishing, 238–241
grinding welds, 237–238

t-stakes, defined, 10
tare weight, 27
task lists, 6
templates
location and relationship marking, 263
making, 146
terms, 10
thoriated tungsten, 56–57
thumbnail dies
defined, 10
shrinking with, 162–165, 169, 189–190
TIG welding
advantages and disadvantages of, 55
aluminum, 43, 46, 58–60, 253
cut line, 274
gases used, 55
grinding beads, 238
patching tips, 289, 291
process, 55–56, 57–58
stainless steel, 60–61
tungsten electrodes, 56–57
tin in solder, 66–67
Tinning Butter, 66, 67, 68–69, 72
tips (cutting)
fuel usage of, 80
importance of size, 80, 85, 86
operation of, 81, 83, 85, 86
swapping welding and, 80
TM Technologies, 44
tools
bead rollers, 18
corking, 10, 13
dolly blocks, 10, 11–12
driving, 10, 13
English wheels, 19
files, 16
hammers, 10–12, 97, 99
keeping track of, 8–9
for making wrinkles, 100
mallets, 15
prying, 13–15
purchasing, 160
quality, 10
shears, 15–16

sheet metal brakes, 15–16
slip rolls, 19–20
spoons and slappers, 10, 12–13
for structural repairs, 7
Vise-Grips, 15, 38, 40, 76–77
welding rods, 38–40
*See also entries beginning with* torches
torch pops, 35
torch-shrinking steel, 223–227
torches (oxyacetylene cutting)
cutting ability of basic, 79
process, 81–83, 84–87
safety, 83, 84
torches (oxyacetylene welding)
filler rods and, 38–40
flange welds, 38
grinding beads, 238
grips, 34, 36
lighting and adjusting, 30–33
making butt joints using, 40–42
manipulating, 35–37
patching tips, 289
shrinking steel, 223–227
shutting down, 34
torches (TIG welding), 55, 58–59
trunk lid repair, 297
tucking
process of cold, 101–107
process of hot, 102, 106, 107–108
tools, 10, 100
tungsten electrodes, 56–57

*The Universal Sheet Metal Machine* (Butler), 166

Victory White Metals, 67
Vise-Grips, 15
flange welds using, 38
making butting joints using, 40
using in riveting, 76–77
voltage, effect of changing, for MIG welders,
50, 53

weld grinding, 237–238
Weld Thru, 292
welding. *See* specific types of welding
welding rods, 38–40
wheel arch, creating shape, 268
wheeling
direction of, 263
pressure while, 271
process, 128–131
tips, 132–134
wire buck building, 180–183
wire edges, making, 142–145
wood bucks, based on Styrofaom models,
183–186
work-hardening, 10, 214, 216
wrinkles, tools for making, 100